WATER IN BIOLOGICAL SYSTEMS

SOSTOYANIE I ROL' VODY V BIOLOGICHESKIKH OB"EKTAKH

СОСТОЯНИЕ И РОЛЬ ВОДЫ В БИОЛОГИЧЕСКИХ ОБЪЕКТАХ

WATER IN BIOLOGICAL SYSTEMS

Volume 1

Edited by L. P. Kayushin

Translated from Russian

cb CONSULTANTS BUREAU · NEW YORK · 1969

First Printing - January 1969
Second Printing - March 1970

Library of Congress Catalog Card Number 69-12513
SBN 306-19001-X

The original Russian edition was published for the Scientific Council on
Problems of Biological Physics of the Academy of Sciences of the USSR by
Nauka Press in Moscow in 1967

FOREWORD

The role of water in biology is perhaps one of the oldest and most extensively discussed problems in natural science. Considering it chronologically, we might begin with Thales, the founder of the Milesian school of philosophy, who considered water to be the basis of everything, including life.

As experimental data subsequently accumulated, water was called upon to an ever greater extent to explain incomprehensible biological phenomena. Although the ideas that have been advanced regarding the role of water in the functioning of biosystems as a whole must often be assigned to the realm of speculative conclusions, there actually are many data for simpler model systems, such as solutions of macromolecules, indubitably indicating that water is to a substantial degree responsible for many of the special properties of macromolecules, not to mention their structure.

The development of new theories of the nature of water and its interaction with matter prompted reexamination of the interpretations of many phenomena pertaining to biosystems. The assumption that many of the properties of water that are now attracting the attention of researchers can actually serve as the key to an understanding of at least some biological processes has become quite prevalent. However, this requires that they be carefully and thoroughly considered by specialists in different fields: physicists, chemists, and biologists.

A symposium convened on the initiative of the Scientific Council on Biophysics of the Academy of Sciences of the USSR in Tbilisi on April 20-25, 1966, was presented with just such a task.

This symposium was the first attempt to summarize the amassed data and the views of different specialists on the problem: "The State and Role of Water in Biosystems."

The papers presented at the symposium fell into three thematic groups.

1. Water and aqueous solutions: surveys of modern theories of water (Yu. V. Gurikov), of aqueous solutions of electrolytes (O. Ya. Samoilov) and nonelectrolytes (G. G. Malenkov), and of the interactions of nonpolar molecules (T. M. Birshtein).

2. Water and macromolecules: data on the interactions of water with macromolecules (P. L. Privalov), on macromolecule hydration (Yu. N. Chirgadze, S. Yu. Veniaminov, S. L. Zimont, G. M. Mrevlishvili, and P. L. Privalov), and on the decisive role of water in the formation of macromolecular structure (N. G. Esipova, Yu. N. Chirgadze, A. A. Vazina, V. V. Lednev, and G. I. Likhtenshtein).

3. Water in biosystems: experimental data on the state of water in biosystems (E. L. Andronikashvili, G. M. Mrevlishvili, P. L. Privalov, A. I. Sidorova, I. N. Kochnev, et al.) and on certain of its functional properties (L. A. Blyumenfel'd, G. V. Fomin et al., G. V. Fomin and V. A. Rimskaya, and Yu. P. Syrnikov).

We hope that the material published in this collection reflects sufficiently well the principal aspects of the problem of water in biosystems, and that it will be interesting and useful to specialists in different fields.

P. L. Privalov

CONTENTS

viii CONTENTS

CURRENT STATE OF THE PROBLEM
OF WATER STRUCTURE

Yu. V. Gurikov

INTRODUCTION

There are many reasons for the increased attention being paid to the problem of water structure. We encounter this compound almost everywhere, but, when one compares it with other liquids, one is struck by the unusual character of its physicochemical properties and molecular structure. Many of the structural details of water have not as yet been determined. The literature has contained reports of puzzling aspects of water, for which it is difficult to find reasonable explanations at our current level of knowledge. We have in mind articles on the ability of liquid water to accelerate many biological processes [1], and on the influence of exposure to magnetic fields on the physicochemical properties of water [2], which is of great economic importance and has found industrial application in a number of cases. Finally, it is a scientifically demonstrated fact that a water column produced in a fine capillary (with a diameter of about 1-10 μ) by condensation from the vapor phase as unusual properties [3, 4].

The rapid development of biology has demonstrated the important role played by water in biological processes. Water is simultaneously the medium for and a direct participant in biochemical reactions. Work by Nemethy and Scheraga [5, 6] has established that water can serve as a connecting link between sections of protein molecules. The stabilization of water structure by nonpolar molecules to form clathrate structures forms the basis for Pauling's molecular theory of anesthesia [7]. All this, as well as the practical demands associated with the search for economical methods of desalting large quantities of sea water, have produced the enormous interest now being displayed in research on water and aqueous solutions.

It must be noted that the study of aqueous solutions is a traditional area of physical chemistry and has a long history. The special significance of water structure is now being constantly emphasized. In speaking of the spatial arrangement of molecules in liquids, it must obviously be remembered that we are dealing merely with average statistical regularities. Framework defects are an inseparable component of water structure, characterizing its erosion as a result of thermal movement of the molecules.

FRAMEWORK STRUCTURE

X-Ray Investigations of Water Structure

The fundamental work of Bernal and Fowler [8] made it clear that the H_2O molecules in both water and ice are tetrahedrally coordinated. Recent investigations of the radial electron-density distribution function [9, 10] confirmed this conclusion and revealed an even closer

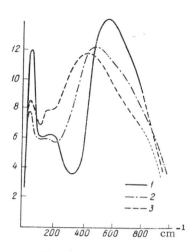

Fig. 1. Phonon spectrum of water [13]. 1) 275°K; 2) 295°K; 3) 365°K.

correspondence between the structures of water and ice. Danford and Levy [10] were able to reproduce the radial distribution function in macromolecules, assuming that the water retains an ice-like framework eroded by thermal movement and having partially filled voids. It was found that, as in the ice structure, each molecule has three neighbors in the same layer, lying at a greater distance (2.94 Å) than the fourth molecule in the adjacent layer (2.77 Å). It must be mentioned that the notion that water retains the short-range ordering of the ice framework and that some of its voids are filled was first advanced by Samoilov in 1946 [11, 12]. It permitted explanation of the increase in the density of water during melting, the density maximum at 4°, and the rise in coordination number with increasing temperature. Almost all recent works on the structure of water and aqueous solutions recognize the existence of voids that can be filled by water molecules.

Scattering of Cold Neutrons

Possibly the most conclusive confirmation of the great similarity of the water and ice structures is the coincidence of the curves representing nonelastic scattering of cold neutrons (with energies of 0.01-0.1 eV) by water at +2° and by ice at −3° [13]. The frequency spectrum retains the same general form as the temperature is raised to 90° (Fig. 1). The two bands (60 and 180 cm^{-1}) belonging to translational vibrations of the molecules as a whole can be ascribed to vibrations of the layers of the ice-like framework with respect to one another and to vibrations of the molecules in the layer planes. The disturbing effect of the molecules in the voids on the force field controlling relative layer movement leads to broadening of the 180 cm^{-1} band, which is especially pronounced in comparison with the constant width of the 60 cm^{-1} band. We can thus say that the similarity of the water and ice structures extends both to the character of molecular coordination and to the fine details of molecular arrangement [14].

Raman Spectrum of Intermolecular Vibrations

The low-frequency region of the Raman spectrum of water was recently studied by Walrafen [15]. This author attributes the decrease in the intensity of the 175 cm^{-1} band with rising temperature to rupture of hydrogen bonds. According to his estimates, 10.5% of the molecules are freed from their bonds. However, this conclusion contradicts the results obtained by Stevenson [16], who found from data on the electronic absorption spectra that the proportion of monomeric molecules in water does not exceed 1% in the most favorable case. Wall and Hornig [17] studied the infrared valence-vibration spectrum of the HDO molecule and concluded that the number of unbonded molecules cannot exceed 5%.

Valence-Vibration Spectrum of the Water Molecule

The assumption that there are two types of hydrogen bonds in water (centrosymmetric within a layer and mirror-symmetric between layers [12, 28]) does not contradict the available data on the infrared valence-vibration spectrum of the H_2O molecule. It must be pointed out that the band in the vicinity of 3000 cm^{-1} has been studied by many researchers, whose explanations are often contradictory. We should call the reader's attention to several recent articles [19, 20] (Fig. 2) in which the valence-vibration spectrum of water is interpreted on the basis of definite structural concepts. Z. A. Gabrichidze detected three maxima, assigning one of

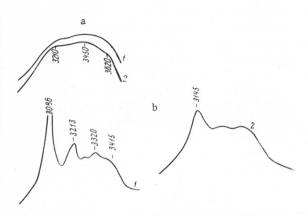

Fig. 2. Valence-vibration spectra of water (a) and ice (b) [19]. Frequency in cm^{-1}. a: 1) 77°C; 2) 27°C; b: 1) 77°K; 2) 260°K.

them (3210 cm^{-1}) to the tetrahedrally coordinated molecules of the ice-like framework, the second (3450 cm^{-1}) to vibrations of the OH groups participating in the bent hydrogen bonds in the deformed framework with filled voids, and the third (3620 cm^{-1}) to the unbonded molecules in the voids. However, strict adherence to Gabrichidze's interpretation forces us to assume that there are many deformed bonds in the ice structure (Fig. 2b). It is difficult to reconcile this with what is known about the ice structure [12]. Another explanation can be advanced on the basis of Gabrichidze's observations [21]. In general, separate maxima in the valence-vibration spectrum should correspond to the two types of bonds. The 3210 cm^{-1} band can be assigned to the stronger, mirror-symmetric vibrations, while the 3450 cm^{-1} band can be ascribed to the longer centrosymmetric bonds.

At low temperatures, the two absorption bands characteristic of ice split into four bands, which correspond to symmetric and asymmetric vibrations of the two types of paired interacting OH-group oscillators. The 3620 cm^{-1} peak apparently corresponds to hydrogen bonds formed by the molecules in the voids.

Dodecahedral Structures

The problem of the structural type of the framework thus cannot be regarded as conclusively solved. As has already been mentioned, the ice framework contains two types of bonds differing in strength and circumjacent symmetry. Hexagonal ice I is distinguished from all frameworks having translational periodicity and tetrahedral molecular coordination by the fact that it contains the maximum number of stronger mirror-symmetric bonds (25%) [18]. Malenkov [22] hypothesized that variants with a larger number of mirror-symmetric bonds than ice can occur in the liquid state, where translational periodicity is lacking. The water molecules in gas-hydrate structures are known to form dodecahedral configurations with a molecular coordination number of four [23]. In contrast to the ice framework, all the bonds in these structures are mirror-symmetric. Pauling [24] worked out his version of the water structure by analogy with clathrates. This model was further developed in an article by Frank and Quist [25], who made a detailed examination of the molecular distribution in the voids of the clathrate framework. There are, however, a number of objections to the Pauling model [14, 26]. As Danford and Levy showed [10], it cannot be reconciled with the radial distribution function of water. Moreover, it is well known that clathrate structures are not formed around polar molecules [27].

Structural Transitions

The possibility of formation of dodecahedral structures of the clathrate type must be considered. Malenkov and Samoilov [28] calculated the electrostatic energy of several structural variants, taking into account the interactions in the first and second coordination spheres. It was found that the energies of the ice-like framework and the clathrate framework are similar, but the ice-like framework is somewhat preferable in terms of energy. However, there are no grounds for denying the possibility that enthalpically disadvantageous structures develop at higher temperatures.

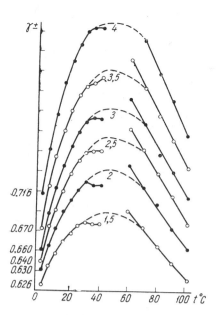

Fig. 3. Activity coefficient of NaCl as a function of temperature and concentration in Drost-Hansen's interpretation [29].

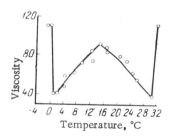

Fig. 4. Viscosity of protoplasm of *Cumingia* (eggs) [30].

The numerous data collected by Drost-Hansen [29, 30] on anomalies in the temperature functions of various physicochemical properties of water and aqueous solutions are of great interest in this connection. Extrema, inflections, and abrupt discontinuities have been noted in density, heat capacity, dielectric constant, viscosity, self-diffusion [31], proton magnetic relaxation, thermal conductivity, magnetic susceptibility, and surface tension. Frontas'ev and Shraiber [32] recently discovered that the polarizability of water molecules undergoes severe changes at 36 and 60°. Similar anomalies have also been observed in the properties of electrolyte solutions (Fig. 3). What was previously regarded as simple scattering of experimental points has been arranged into a coherent, orderly system that indicates a sharp change in molecular ordering over the range 30-60°. Drost-Hansen [29] found four characteristic temperatures (15, 30, 45, and 60°) near which there are abrupt changes in the properties of water. The presence of solutes has little influence on the transition temperatures. Drost-Hansen [29] assumes that qualitative structural rearrangements take place in water at the aforementioned temperatures. It is at present difficult to establish the nature of these transformations with any degree of accuracy. One possible interpretation suggests itself: conversion of the ice-like framework to dodecahedral structures of the clathrate type and perhaps to structures characteristic of other ice modifications [14]. The great diversity of clathrate structures and ices supports this theory.

Drost-Hansen [30] demonstrated that the existence of discontinuities in the properties of water is convincingly manifested in biological processes. Figure 4 shows just one curious example. The abrupt changes in this case are associated with temperatures of 15 and 30°. We do not as yet have the necessary knowledge and experimental data, so that it is difficult to comprehend the true nature of these phenomena. It has become obvious, however, that knowledge of water structure is of exceptional importance for proper interpretation of the mechanisms of biological processes.

FRAMEWORK DYNAMICS

Nature of Intermolecular Interactions in Water

Any description of the structure of liquid water would be incomplete if nothing were said about the defects, or disturbances, in the ordered arrangement of the molecules. While the number of defects in ice is very small, these disturbances become an important structural element in the liquid state. Appearance of defects cannot but be accompanied by bending and even complete rupture of the bonds, since it would otherwise be impossible to imagine self-diffusion

and other transfer processes [33-35]: viscous flow, dielectric relaxation, and electrical conductivity.

The orientation of adjacent molecules can vary within wide limits in the electrostatic
model of the hydrogen bond, the bonds merely being weakened in this case. The notion of the
bending of hydrogen bonds was proposed by Pople [36] on the basis of a quantitative theory of
water structure. It is important to emphasize that all the molecules and all the bonds are
equivalent in the Pople model. There is only a slight scattering about the mean, so that we can
speak of a diffuse energy band. Although, as Sokolov showed [37], the electrostatic model is
essentially unable to explain the principal spectroscopic manifestation of the hydrogen bond
(the displacement of the valence-vibration frequency of the OH group into the long-wave region),
the notion of the existence of diffuse energy bands seems very promising for further development of the theory of water structure. The first step in this direction was recently taken by
Vand and Senior [38], who considered a model with three energy bands corresponding to molecules whose protons participate in one or two hydrogen bonds or are not bound at all. On the
other hand, if we assume that the hydrogen bond has an admixture of covalency, only slight deviations of the proton from the bond angle are permissible. Large angular deviations are
equivalent to complete rupture of the bond. Proceeding from the concepts of the donor−accep-
tor character of the hydrogen bond developed by Sokolov [37], Frank [39] evolved extremely important and fruitful hypotheses regarding the cooperative nature of the development and breakdown of the hydrogen-bond system in water. According to the covalent theory of the hydrogen
bond, transfer of unshared electrons from the oxygen atom of one molecule to unoccupied orbit-
als of a hydrogen atom in an adjacent molecule is accompanied by charge separation. In this
case, one of the interacting molecules takes on an excess positive charge and becomes "acidic"
in character, while the other takes on a negative charge and acquires "basic" properties. As a
result, it turns out that molecules joined by a hydrogen bond are capable of forming stronger
bonds to other adjacent molecules. We can therefore speak of the simultaneous formation or
disappearance of a large number of hydrogen bonds. Frank applied these considerations to a
model of "flickering clusters," i.e., unstable molecular associations capable of undergoing
structural rearrangement over a period of about 10^{-10}-10^{-11} sec. Using this model, he was
able to explain the similarity of the activation energies of self-diffusion, viscosity, and dielectric relaxation (4.0-4.5 kcal/mole).

Mosaic Models

Frank's model is among the two-structure mosaic models, in which water is treated as
an ideal mixture of segments having an ice-like structure and segments lacking hydrogen bonds
and having a more compact molecular arrangement. This model was refined by Nemethy and
Scheraga [5, 27, 40], whose work had a great influence on the subsequent development of theoretical research on the structure and thermodynamic properties of water. The principal value
of their method lies in the fact that they were able to find a model procedure for evaluating the
state of the group surface and for calculating the number of molecules in a group. However, a
number of objections have been raised to the Frank−Nemethy−Scheraga model [26, 38, 41]. It
is difficult to reconcile it with the data on the infrared and Raman spectra of water in the
valence-vibration region. According to the estimate made by Wall and Hornig [17], the number
of molecules incorporated into ice-like structures does not exceed 5%. It is also hard to bring
mosaic models into agreement with the large fluctuations in coordination number observed by
Fisher et al. [43, 44].

In discussing mosaic models, we should not fail to mention a recent version proposed by
Davis and Litovitz [45]. These authors distinguished two six-member rings located one above
the other in the ice-like framework. If one of them is rotated by 60°, the rings become more

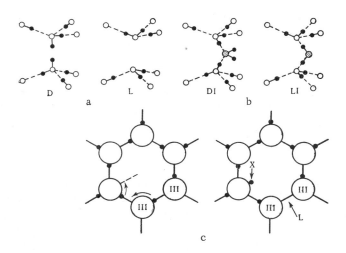

Fig. 5. Defects in the ice framework. a) D and L represent
Bjerrum defects [8]; b) DI and LI represent Haas defects
[50], with I indicating a molecule in a framework void; c) X
represents a Dunitz defect [51], the proton of the molecule
being directed toward the center of the framework void.

tightly packed and form a body-centered cubic lattice. The molecules in this model are more
rigidly bonded than in Frank's flickering-cluster model. It is totally unclear how the fact that
the self-diffusion coefficient for water is 100,000 times that for ice can be interpreted within
the framework of this model [46].

Two-State Models

At least partial resolution of the contradictions described above has apparently been
achieved in two-structure models of another type: two-state models. The molecules at the
junctions of the three-dimensional hydrogen-bond network correspond to one state, while the
molecules occupying the framework voids correspond to the other. Two-state models have
been considered by Frank and Quist [25] and by Marchi and Eyring [47] for the clathrate framework. Samoilov's model (an ice-like framework with filled voids) must be regarded as the
physical basis for Hall's abstractly derived two-structure model [48]. Fisher and Samoilov
[49] attribute the high mobility of molecules in water to rapid movement through the framework
voids.

The lack of optical-spectrum bands that could be ascribed to monomeric molecules
forces us to acknowledge that the molecules in the voids can form hydrogen bonds with the
framework molecules.* These are obviously very weak as a result of steric difficulties. We
do not at present have complete information on the state of the molecules in the voids. Haas
[50] discussed one possible model (Fig. 5). Dunitz [51] cited considerations indicating high stability of the X-defects (Fig. 5). Gurikov [46] demonstrated that self-diffusion in both liquid
water and ice is effected through Brownian movement of the DI- and LI-defects, which is accompanied by rapid exchange of framework molecules for the molecules in the voids.

---

*This bond is anomalously long, about 3.1 Å. It will be remembered that the normal length of
  an O−H−O hydrogen bond in crystalline materials usually ranges from 2.45 to 2.95 Å.

It proved possible to obtain some information on the state of the molecules in the voids by calculating the thermodynamic properties of water for a generalized Samoilov model, using an ice-like framework containing DI- and LI-defects [26, 52, 53]. Forming hydrogen bonds with the framework molecules, the void molecules act to destroy the framework. Calculations have shown [52] that bonds in which the void molecules participate are thermodynamically more favorable than the framework bonds. This results from the fact that rupture of the framework bonds leads to a large rise in entropy, which is superimposed on the energy loss during bond rupture.

What consequences can this have for the structure of water? The reader will recall Frank's proposed cooperative mechanism for the formation and breakdown of hydrogen bonds. It is clear that the bond-rupture process, having begun somewhere near a filled void, encompasses large areas of the framework. Calculations have shown that the greater the extent to which the framework is destroyed, the larger is the number of bonds formed by the void molecules and the greater is the gain in free energy during filling of new voids. Hence it follows that large areas in the ice-like framework are broken up and the number of filled voids is increased and formed near filled voids. An "association" of the void molecules occurs. Thus, in contrast to the Frank—Nemethy—Scheraga model, water must be regarded as having a "shaky" ice-like framework, in which regions having a more compact but orientationally disordered structure have melted out.

The existence of a "conglomeration" of void molecules permits a qualitative interpretation of the minimum in the heat capacity C_p at about 35°. The initial decrease in heat capacity as the temperature is raised from 0 to 30° apparently results from dominance of processes accompanied by hydrogen-bond rupture. The expenditure of heat in bond rupture decreases over the entire deformed framework. At temperatures above 40°, bond rupture is succeeded by another process that leads to some structural stabilization and is apparently caused by development of molecular associations, whose breakdown requires an additional quantity of heat. These associations are probably aggregates of void molecules in the deformed ice-like framework; they increase in number as the three-dimensional hydrogen-bond network becomes generally weaker [54, 55].

LITERATURE CITED

1.   A. K. Guman, Tr. Leningr. Obshch. Estestvoispytat., 70:70 (1959).
2.   V. I. Klassen, Dokl. Akad. Nauk SSSR, 166(6):1383 (1966).
3.   B. V. Deryagin and N. N. Fedyakin, Dokl. Akad. Nauk SSSR, 146(2):403 (1962).
4.   B. V. Deryagin, M. V. Talaev, and N. N. Fedyakin, Dokl. Akad. Nauk SSSR, 165(3):597 (1965).
5.   G. Nemethy and H. A. Scheraga, J. Chem. Phys., 36:3401 (1962).
6.   G. Nemethy and H. A. Scheraga, J. Phys. Chem., 66:1773 (1962).
7.   L. Pauling, Science, 134:15 (1961).
8.   J. D. Bernal and R. H. Fowler, J. Chem. Phys., 31:515 (1933).
9.   G. W. Brady and W. J. Romanow, J. Chem. Phys., 32:306 (1960).
10.   M. D. Danford and H. A. Levy, J. Am. Chem. Soc., 84:3965 (1962).
11.   O. Ya. Samoilov, Zh. Fiz. Khim., 20(12):1411 (1946).
12.   O. Ya. Samoilov, Structure of Aqueous Electrolyte Solutions and the Hydration of Ions, Consultants Bureau, New York (1965).
13.   K. E. Larsson and V. Dahlborg, J. Nucl. Energy, 16(2):81 (1962).
14.   Yu. V. Gurikov, Zh. Strukt. Khim., 4(6):824 (1963).
15.   G. E. Walrafen, J. Chem. Phys., 40(11):3249 (1964).
16.   D. P. Stevenson, J. Phys. Chem., 69(7):2145 (1965).
17.   T. T. Wall and D. F. Hornig, J. Chem. Phys., 43(6):2079 (1965).

18. N. Bjerrum, Dan. Mat. Phys. Medd., Vol. 27, No. 1 (1951).
19. Z. A. Gabrichidze, Optika i Spektroskopiya, 19(4):575 (1965).
20. Z. A. Gabrichidze, in: State of Water in Living Organisms [in Russian], Izd. LGU (1966).
21. Yu. V. Gurikov, in: State of Water in Living Organisms [in Russian], Izd. LGU (1966).
22. G. G. Malenkov, Dokl. Akad. Nauk SSSR, 137(6):1354 (1961).
23. G. G. Malenkov, Zh. Strukt. Khim., 3:220 (1962).
24. L. Pauling, The Hydrogen Bond, London (1959).
25. H. S. Frank and A. S. Quist, J. Chem. Phys., 34(2):604 (1961).
26. Yu. V. Gurikov, Zh. Strukt. Khim., 6(6):817 (1965).
27. G. Nemethy and H. A. Scheraga, J. Chem. Phys., 36(12):3382 (1962).
28. G. G. Malenkov and O. Ya. Samoilov, Zh. Strukt. Khim., 6(1):9 (1965).
29. W. Drost-Hansen, First International Symposium on Water Desalination, Washington, October, 1965.
30. W. Drost-Hansen, in: Symposium, "Forms of Water in Biological Systems," New York Academy of Sciences, October, 1964.
31. T. H. Wang, J. Am. Chem. Soc., 73:4181 (1951).
32. V. P. Frontas'ev and L. S. Shraiber, Zh. Strukt. Khim., 6(4):512 (1965).
33. T. H. Wang, J. Phys. Chem., 58:683 (1954).
34. G. H. Haggis, J. B. Hasted, and T. J. Buchanan, J. Chem. Phys., 20(9):1452 (1952).
35. R. A. Horne and R. A. Courant, J. Phys. Chem., 6(7):2224 (1965).
36. J. A. Pople, Proc. Roy. Soc. London, A202:323 (1950).
37. N. D. Sokolov, Usp. Fiz. Nauk, 57:205 (1955).
38. V. Vand and W. A. Senior, J. Chem. Phys., 43(6):1878 (1965).
39. H. S. Frank, Proc. Roy. Soc. London, A247:481 (1958).
40. G. Nemethy and H. A. Scheraga, J. Chem. Phys., 41:680 (1964).
41. O. Ya. Samoilov and T. A. Nosova, Zh. Strukt. Khim., 6(4):798 (1965).
42. V. Vand and W. A. Senior, J. Chem. Phys., 43(6):1869, 1873 (1965).
43. I. Z. Fisher and V. K. Prokhorenko, Dokl. Akad. Nauk SSSR, 123:131 (1958).
44. I. Z. Fisher and V. I. Adamovich, Zh. Strukt. Khim., 4(6):819 (1963).
45. C. M. Davis and T. A. Litovitz, J. Chem. Phys., 2:2563 (1965).
46. Yu. V. Gurikov, Zh. Strukt. Khim., 5(2):188 (1964).
47. R. P. Marchi and H. Eyring, J. Phys. Chem., 68(2):221 (1964).
48. L. Hall, Phys. Rev., 73(7):775 (1948).
49. V. K. Prokhorenko, O. Ya. Samoilov, and I. Z. Fisher, Dokl. Akad. Nauk SSSR, 125:356 (1959).
50. C. Haas, Phys. Letters, 3:126 (1962).
51. J. D. Dunitz, Nature, 197:860 (1963).
52. Yu. V. Gurikov, Zh. Strukt. Khim., 7(1):8 (1966).
53. V. M. Vdovenko, Yu. V. Gurikov, and E. K. Legin, in: State of Water in Living Organisms [in Russian], Izd. LGU (1966).
54. V. M. Vdovenko, Yu. V. Gurikov, and E. K. Legin, Zh. Strukt. Khim., Vol. 7, No. 4 (1966).
55. V. M. Vdovenko, Yu. V. Gurikov, and E. K. Legin, Zh. Strukt. Khim., Vol. 7, No. 6 (1966).

HYDROPHOBIC INTERACTIONS OF NONPOLAR MOLECULES

T. M. Birshtein
Institute of High-Molecular Compounds
Academy of Sciences of the USSR

An important characteristic of biopolymer molecules is the fact that they contain compact intramolecular structures in solution. The compactness can be one-dimensional (the α-helix or Crick–Watson helix), two-dimensional (the β-structure in polypeptides), or three-dimensional (globular proteins) in character. These structures are stabilized by noncovalent interactions; a change in the ambient conditions (temperature, pH, or solvent composition) leads to cooperative breakdown of the compact structures.

In order to characterize the interaction forces within the structure, we can consider the distance between pairs of atoms that are not valently bonded. This distance should be 2-3 Å if there is a hydrogen bond or salt linkage in the pair; if the pair is bonded by Van der Waals forces, the distance should be of the order of the sum of the Van der Waals radii, i.e., 3-4 Å. When all the pairs of atoms separated by short distances in myoglobin were printed out by a computer, the pairs of the first type comprised a column 0.25 inch long, while those of the second type occupied a column 2 inches long, i.e., there were eight times as many of the latter [1]. It should be noted that this situation obtains in a protein with a large helix content (70% α-helix). The DNA double helix is another example. Each pair of bases in this helix contains two or three pairs of atoms separated by a distance corresponding to the hydrogen bond and (counting the atoms in the heterocyclic structures) 15 pairs of atoms separated by the Van der Walls-contact distance, i.e., there are 5-6 times as many Van der Waals pairs as hydrogen-bonded pairs.

It must be emphasized that the bipolymer molecules that have been studied were in solution, i.e., in water; there would thus also be atomic pairs (polymer–solvent pairs) at the Van der Waals distances in the case of a loose, noncompact macromolecular structure.* In considering the interactions within the macromolecule, we must consequently take into account the influence exerted on them by the solvent, in the sense of competition between the polymer–polymer (and hence solvent–solvent) and polymer–solvent interactions. Interactions of all the aforementioned types are also the principal factor governing the solubility of compounds. If the solute molecules consist of groups of the same type, they dissolve in a given solvent if formation of a solvent–solute contact rather than solvent–solvent and solute–solute contacts

---

*Such pairs cannot be detected by x-ray diffraction analysis if, as is usually the case, the water molecules are not rigidly bound to the polymer.

Table 1. Free Energy, Energy, and Entropy of Migration
of Hydrocarbon from Nonpolar Environment to Water at 25°*

Compound	ΔF_{tr}, kcal/mole	$\Delta H^!_{tr}$, kcal/mole	ΔS_{tr}, e.u.
CH_4	+3.0	−2.6	−13
C_2H_0	+3.8	−2.0	−19
C_3H_8	+5.0	−1.8	−23
C_4H_{10}	+5.8	−1.0	−23
C_6H_6	+4.3	$0 \sim +0.6$	−14
$C_6H_5{-}CH_3$	+5.0	$0 \sim +0.6$	−16

*Hydrocarbon in nonpolar solvent $\overset{K_{tr}}{\rightleftharpoons}$ hydrocarbon in water:

$$\Delta F_{tr} = -\kappa T \ln K_{tr} ;$$

liquid hydrocarbon $\overset{K_{tr}}{\rightleftharpoons}$ hydrocarbon in water:

$$\Delta F_{tr} = -\kappa T (\ln K_{tr} - \ln x)$$

where x is the molar concentration.

leads to a decrease in free energy.* This almost always occurs when the solvent and solute are of the same type (like dissolves in like). In the case of macromolecules consisting of groups of the same type, the favorability of polymer−solvent contacts necessary for solubility causes a macromolecule in solution to take the form of a loose statistical cluster arrayed in a volume that substantially exceeds that of a molecule of dry polymer.

Another situation obtains when the solute consists of groups of different types, e.g., polar and nonpolar. The polar groups can provide solubility if the solvent is polar, while the tendency of the nonpolar groups to go out of solution is manifested in their mutual attraction and can, under appropriate conditions, cause them to aggregate.

The interactions responsible for the insolubility of a compound in a given solvent are customarily referred to as lyophobic interactions [2], generalizing the term hydrophobic interactions, which was proposed by Kauzmann [3], to describe the relative attraction of nonpolar groups in an aqueous medium.

It is obvious that, since there are molecules and groups that are soluble and insoluble in every solvent, one can theoretically always find a system soluble in a given solvent that will contain soluble and insoluble groups and in which lyophobic interactions will occur. Water has no special features in this respect, but such features can be manifested in the physical nature and magnitude of the effective forces, their dependence on various factors (temperature and pressure), etc. We will henceforth consider the hydrophobic interaction of nonpolar groups in water.

As has already been emphasized, this interaction results both from the interaction of the groups with one another and from their interaction with the water, which is to a substantial extent governed by the structure of the latter. We can employ the free energy of migration of one mole of a nonpolar compound from a nonpolar environment (a solution in a nonpolar solvent or a liquid hydrocarbon) to water as a measure of the hydrophobic interaction. Table 1, which was compiled from articles by Kauzmann [3] and Nemethy and Scheraga [4], gives values for this

---

*More precisely, since solution processes occur at constant pressure, we should speak not of free energy and energy but of thermodynamic potential and enthalpy, respectively. Since the term pV is small, these quantities are almost identical in our case.

factor and for the heat and entropy of migration found from its temperature function. Plus signs before the free energy and energy indicate that solution is disadvantageous. It can be seen that solution of aliphatic hydrocarbons is advantageous in terms of energy, the positive values of ΔF being associated with the entropy term.* The energy term for aromatic hydrocarbons at room temperature is zero or slightly positive, and the disadvantage of solution also has an entropic character. These thermodynamic characteristics of the hydrophobic interaction result from the specific properties of water.

It is well known that water has a number of characteristics distinguishing it from ordinary liquids. Specifically, its density increases rather than decreases on melting and passes through a maximum at 4°, after which it falls as in ordinary liquids. The increase in the density of water on melting is possible because, as can be seen from the distance between the closest molecules in the ice structure, the molecular radius is 1.4 Å and the closest packing corresponds to a density of 1.8 g/cm^3. Hence it follows that the ice structure is extremely open; actually, the coordination number for ice is 4. This is due to the fact that each water molecule is capable of forming four hydrogen bonds (two as a donor and two as an acceptor). The coordination number increases somewhat during melting and then rises with the temperature. The presence of a maximum in the density−temperature curve shows (as a minimum and maximum in any function always show) that there are two temperature processes proceeding in different directions. One is the decrease in density with rising temperature as a result of the increase in disordering, which is common to all liquids. It is opposed by another process, which produces an increase in density with rising temperature and is dominant at low temperatures.

There is as yet no single physical picture that would permit us to describe all the properties of water and aqueous solutions with sufficiently reliable premises and a sufficiently rigorous theoretical apparatus, although the first detailed article on water structure [5] appeared more than 30 years ago. Nevertheless, the discrepancies among various models are not all that fundamental. On the whole, water is a highly structured liquid containing a substantial proportion of an open tetrahedral structure of low density and a corresponding ice structure. There are also areas of high density produced by rupture of hydrogen bonds; it is the appearance of these areas that causes the increase in the density of water on melting.

The two structures in water differ greatly in their structural and energy parameters. The energy, entropy, density, mobility, distance to closest adjacent molecule, and coordination number are lower for the open than for the compact structure.

The differences among the models proposed by different authors consist in part in somewhat differing concepts of the structure of the low-density areas (the character of the lattice), but they pertain principally to the compact structural areas.

At this point, we can distinguish two groups of models. The first group, proposed by Frank and Wen [6, 7], takes into account the fact that hydrogen bonds are formed cooperatively as a result of the change in electron density during formation of a single hydrogen bond. The hydrogen-bonded segments are therefore flickering clusters surrounded by an amorphous liquid. According to the second group of models, nonhydrogen-bonded or incompletely bonded molecules are found in the lattice voids (the model proposed by Samoilov [8], Pauling [9], and Frank and Quist [10]). These models do not differ as greatly as would appear at first glance (compare [11]). The first model (flickering clusters) incorporates two basic states: molecules with four hydrogen bonds in the cluster and molecules without hydrogen bonds in the dense amorphous

_____

*This means that, for solution of liquid or gaseous hydrocarbons, the total entropy of solution is less than the entropy of mixing.

liquid. In addition, more detailed thermodynamic theories [12] consider molecules with one, two, and three hydrogen bonds in the boundary regions. On the other hand, the theories that provide for interstitial molecules in the voids use the framework and void molecules as the two basic states. In actuality, the presence of molecules in the voids distorts the lattice structure and such segments are not equivalent to areas of the lattice where there are no interstitial molecules. All the molecules in the defective regions have properties similar to those of free water. It can therefore be said that the two sets of models are different thermodynamic approaches to the same complex structure, in which the initial averaging is carried out for different classes of molecules. Experimentation has not made it possible to choose among the models. According to x-ray data, intermolecular distances of about 2.8 and 4.5 Å, which correspond to the ice structure, and a whole set of intermediate distances are observed in water. This can correspond either to models using molecules in the voids of an open structure or models using dense amorphous water. Proceeding from data on the infrared spectra of water, Senior and Vand [15] recently showed that the spectroscopic and thermodynamic properties of water can be described by a single model only if the energy level blurring is taken into account, i.e., if the energy spectrum contains comparatively broad (about 1 kcal/mole) bands. Although there are thermodynamic theories of water for the different models, the most detailed calculations for hydrophobic interactions were made by Nemethy and Scheraga [4] for the flickering-cluster model. According to their work, the average number of molecules in a cluster decreases from 90 to 28 when the temperature is raised from 0 to 60°, while the percentage of unbroken hydrogen bonds drops from 53 to 37%. The unbonded water molecules occupy one or two layers near the cluster.

Let us now turn to data on the energies of solution of nonpolar compounds in aqueous solutions. As has already been noted, the high positive free energy (low solubility) of these compounds results from the large decrease in entropy during solution, while the enthalpy favors solution for aliphatic hydrocarbons and is almost zero (at room temperature) for aromatic hydrocarbons. According to the hypothesis advanced by Frank and Evans [16], the large negative entropy of solution means that placing a hydrocarbon in water leads to an increase in the structuring of the latter, i.e., causes a shift in the equilibrium toward an open ice-like structure having lower entropy. This is understandable in light of the fact that the ice-like structure is very loosely packed, contains large pores, and is characterized by a low coordination number. The hydrocarbon molecules, not being at all similar to the water molecules, cannot be incorporated into this structure, but they can fit into its pores, at least near its boundaries. This causes an increase in coordination number for the water molecules adjacent to the hydrocarbon, i.e., there is a gain in the energy of the hydrocarbon−water Van der Waals contact, while the water−water contacts persist and only the hydrocarbon−hydrocarbon contact is broken. At the same time, localization of a hydrocarbon in a dense-structure region leads not to additional contacts (the structure remains equally dense), but merely to replacement of the water−water contact by an energetically less advantageous water−hydrocarbon contact (the dipole−dipole interaction is replaced by a dispersion interaction). During solution of a hydrocarbon in water, it is selectively dissolved in the structured liquid, which displaces the equilibrium toward greater structuring, i.e., leads to a decrease in entropy.

The presence of ordered water around nonpolar compounds is confirmed by the existence of crystalline hydrates of a number of nonpolar gases and liquids. The water molecules in these hydrates are arranged in such fashion as to form voids 5-7 Å in diameter. The structure is stabilized by inclusion of solute molecules of appropriate size in the voids. This does not mean that nonpolar compounds rigidly stabilize a given structure in solution, but it does mean a shift in equilibrium in this direction. In contrast to crystalline hydrates, there are no complete hydrate cells in solution, both for steric reasons and because the cooperative structural stabilization that takes place in crystalline hydrates does not occur in dilute solutions. The

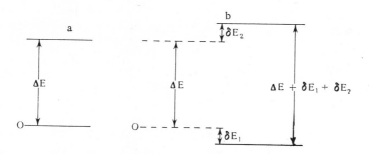

Fig. 1. Energy levels in two-structure water model.
a) In absence of dissolved hydrocarbon; b) for water
molecules near dissolved hydrocarbon molecule.

decrease in entropy resulting from the increase in the structuring of water on solution of hydrocarbons is also responsible for the low solubility of these compounds.

Let us illustrate the foregoing with calculations for a simple model. If water can have two molecular states, corresponding to the ordered open structure and the disordered compact structure, the average fraction of the water molecules in the open structure ϑ_a^0 is

$$\vartheta_a^0 = \frac{1}{1 + K^0},\tag{1}$$

where K^0 is the equilibrium constant for the two states. The superscript 0 indicates that we are considering the water in the absence of a solute. The temperature function of the equilibrium constant

$$K^0 = e^{-\frac{\Delta\widetilde{E} - T\Delta\widetilde{S}}{\kappa T}}\tag{2}$$

for cooperative systems is governed not by the magnitude ΔE of the energy difference between the water molecules in the two states (Fig. 1) but by the factor $\Delta\widetilde{E}$, which is proportional to it and represents the effective energy difference between the two structures per cooperative unit (compare [17]). In similar fashion, $\Delta\widetilde{S}$ is the effective entropy difference between the two structures. Experimental data have shown that the structuring of water decreases with rising temperature, i.e., $\Delta\widetilde{E} > 0$, and K_0 increases with temperature.

When a hydrocarbon molecule is placed in water, the energy of the adjacent water molecules is altered; since the hydrocarbon−water dispersion interaction weakens as the sixth power of the intermolecular distance, only the energy of the water molecules in the first coordination sphere around the hydrocarbon molecule changes. As has already been pointed out, the energy of a water molecule in the open structure decreases by an amount δE_1 (Fig. 1), which is the energy of the hydrocarbon−water Van der Waals contact ($\delta E_1 > 0$), while the energy of a molecule in the compact structure increases by δE_2, which is the difference in the energies of the hydrocarbon−water and water−water interactions. The average fraction of the water molecules in the open structure (the degree of structuring) near the hydrocarbon molecule is

$$\vartheta_a = \frac{1}{1 + K},\tag{3}$$

where K is the equilibrium constant for the two structures in the water layer around the hydrocarbon molecule. Since the distance between the energy levels of the two structures increases for water molecules near the hydrocarbon molecule, $\Delta E + \delta E_1 + \delta E_2 > \Delta E$, we have $K < K^0$

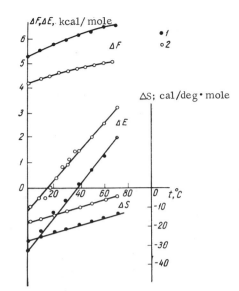

Fig. 2. Free energies, energies, and entropies of migration of hydrocarbons from nonpolar environment to water. 1) Butane; 2) benzene (results yielded by processing of experimental data on solubility of hydrocarbons [18-22] with equation [4]).

and $\vartheta_a > \vartheta_a^0$, i.e., the structuring of the water increases in the layer near the hydrocarbon; the change in the fraction of the water molecules in the open structure is

$$\Delta\vartheta_a = \vartheta_a - \vartheta_a^0 = \frac{1}{1+K} - \frac{1}{1+K^\circ} = \frac{\vartheta_a^0\,(1-\vartheta_a^0)\left(1-\dfrac{K}{K^0}\right)}{\vartheta_a^0 + (1-\vartheta_a^0)\,\dfrac{K}{K^0}}. \quad (4)$$

If the coordination number of a hydrocarbon molecule in water equals X, X $\cdot \Delta\vartheta_a$ water molecules pass from the disordered compact structure to the ordered open structure during solution of each hydrocarbon molecule.

The ratio K/K^0 for a noncooperative system would equal $e^{-(\delta E_1 + \delta E_2)/kT}$, while the values of δE_1 and δE_2 are governed by the comparatively small energies of the hydrocarbon−water and water−water interactions in the dense structure. We can therefore substitute $K/K^0 \simeq 1$ into the denominator of Eq. (4), obtaining

$$\Delta\vartheta_a = \vartheta_a^0(1-\vartheta_a^0)\left(1-\frac{K}{K^0}\right) \quad (5)$$

each dissolved hydrocarbon molecule therefore causes transfer of less than X $\cdot \vartheta_a^0$ $(1-\vartheta_a^0)$ water molecules to the open structure. The fact that the proportion of the open structure in pure water ϑ_a^0 ranges from 30 to 70% means that solution of a hydrocarbon molecule in water leads to structuring of less than X/4 water molecules. According to the estimates of Nemethy and Scheraga [4], the coordination numbers of the first few members of the simple aliphatic−hydrocarbon series (from methane to octane) and of benzene and its derivatives lie within the range X = 15-30. During solution of each of these molecules, several water molecules (less than 4-8) accordingly pass from the dense disordered structure into the open structure. Similar estimates made from Eq. (4) are also valid for a real cooperative system.

The number of water molecules migrating to the open structure is a function of temperature, since ϑ_a^0 and K/K^0 depend on temperature. According to the estimates made by Nemethy and Scheraga [4], the structuring of the water decreases from 50 to 30% as the temperature is raised from 0 to 70°. In this case, K/K^0 increases and the number of molecules migrating to the open structure X $\cdot \Delta\vartheta_a$ therefore decreases with rising temperature.

Since the entropy of the water molecules in the open structure is less than that of the molecules in the dense structure, the migration of water molecules to the open structure during solution of a hydrocarbon is associated with a loss of entropy. Precise calculation of the entropy of the system requires a more or less detailed description of the character of its cooperativity. However, we can draw some conclusions from the foregoing discussion without making this calculation. The proportionality of the number of water molecules migrating to the open structure during solution of a hydrocarbon molecule to the coordination number shows that the decrease in entropy that occurs when a hydrocarbon molecule is placed in water should increase with the size of the molecule. The drop in $\Delta\vartheta_a$ with rising temperature indicates that the loss of entropy during solution should decrease as the temperature increases. Both these qualitative conclusions are in agreement with experimental data (see Fig. 2). The loss of entropy during rearrangement of the water structure is also responsible for the high positive free energy of the migration of hydrocarbons to water (Table 1).

The change in the energy levels of water molecules and their redistribution over these levels during solution of a hydrocarbon leads to a change in their average energy. The energy of a water molecule in the absence of a hydrocarbon (the energy of a molecule in the open structure is taken as zero) equals

$$E_0 = (1 - \vartheta_a^0)\,\Delta E = \frac{K^0}{1 + K^0}\,\Delta E.$$

The average energy of a water molecule near a hydrocarbon molecule is

$$E = -\delta E_1\,\vartheta_a + (\Delta E + \delta E_2)(1 - \vartheta_a), \tag{7}$$

while the total energy of migration of a hydrocarbon molecule from a nonpolar environment to water is

$$E_{tr'} = E_n + X\,(E - E_0), \tag{8}$$

where E_n is the energy of the interaction of a hydrocarbon molecule with all the molecules surrounding it in the nonpolar medium ($-E_n < 0$),

$$E - E_0 = \delta E_2 - (\delta E_1 + \delta E_2)\,\vartheta_a - \Delta E \cdot \Delta\vartheta_a, \tag{9}$$

and ϑ_a and $\Delta\vartheta_a$ are determined from Eqs. (3) and (4).

It follows from Eqs. (8) and (9) that the energy of migration of a hydrocarbon from a nonpolar environment to water depends both on the energies of the hydrocarbon—hydrocarbon and hydrocarbon—water contacts and on the decrease in the energy of the water molecules resulting from structuring, i.e., the increase in the proportion of the open structure during solution of the hydrocarbon. Since ϑ_a and $\Delta\vartheta_a$ decrease with rising temperature, the energy of migration of a hydrocarbon to water should increase (in the algebraic sense) with temperature.

We showed above that the loss of entropy during solution of hydrocarbons drops with rising temperature. The decrease in entropy is due to the dependence on temperature of the proportion of structured water molecules $\Delta\vartheta_a$, while the increase in the energy of solution results from this factor, from the increase in the fraction $(1 - \vartheta_a)$ of the water molecules in the dense structure (whose energy rises by δE_2 during solution of a hydrocarbon), and from the decrease in the fraction ϑ_a of the molecules in the open structure, whose energy decreases by δE_1 during solution of a hydrocarbon. The increase in the energy of solution with temperature should thus increase as the loss of entropy decreases, and the free energy of migration of a hydrocarbon to water should consequently rise with temperature. These conclusions are in agreement with experimental data (Fig. 2).

It must be noted that direct calculation of the free energy of migration of a hydrocarbon from a nonpolar environment to water requires more detailed assumptions regarding the character of the cooperativity in the system than does calculation of the energy of migration. If we assume that the interaction of water molecules does not have a cooperative character, the statistical sum of the water molecules in the absence of a hydrocarbon is

$$Z_0 = 1 + e^{-\frac{\Delta E - T\Delta S}{kT}}, \tag{10}$$

where ΔE and ΔS are the differences in the energies and entropies of the dense and open structures, the energy and entropy of the latter structure being taken as zero. The statistical sum of the water molecules near a dissolved hydrocarbon is

$$Z = e^{+\frac{\delta E_1}{kT}} + e^{-\frac{\Delta E + \delta E_2 - T\Delta S}{kT}}, \tag{11}$$

where the difference between the first term and unity is governed by the decrease in the energy of a water molecule in the open structure near the hydrocarbon. The contribution made by the water molecules to the free energy of migration of a hydrocarbon molecule from a nonpolar environment to water equals

$$\Delta F = -X k T \ln \frac{Z}{Z_0} = -X \delta E_1 - XkT \ln \frac{1+K}{1+K^0} = -X \delta E_1 + XkT \ln \frac{\vartheta_a}{\vartheta_a^0}. \tag{12}$$

In deriving Eq. (12), we took into account the fact that, for a noncooperative system, the effective energy and entropy of the cooperative unit in Eq. (2) reduce to the corresponding figures for a single molecule.

The temperature function of the energy of migration of a hydrocarbon from a nonpolar environment to water in Eqs. (8) and (9) is governed by the quantities ΔE and $\delta E_1 + \delta E_2$. Both these factors are determined solely by the energy of the water−molecule interaction and are not associated with the energy of the hydrocarbon−water interaction. This is obvious for ΔE; as for $\delta E_1 + \delta E_2$, δE_1 is the energy of the hydrocarbon−water interaction and δE_2 is the difference in the energies of the water−water and hydrocarbon−water interactions, so that dependence on the energy of the hydrocarbon−water interaction is excluded in the sum $\delta E_1 + \delta E_2$. The temperature-dependent portion of E_{tr} is accordingly independent of the structure of the dissolved hydrocarbon, so far as the model under consideration is valid; specifically, it should be the same for aromatic and aliphatic hydrocarbons, although the values of δE_1 and δE_2 differ substantially for these compounds. Thus, according to the estimates made by Nemety and Scheraga [4], $\delta E_1 = 0.03$ and $\delta E_2 = 0.30$ kcal/mole for aliphatic hydrocarbons, while $\delta E_1 = 0.16$ and $\delta E_2 = 0.18$ kcal/mole for aromatic hydrocarbons; $\delta E_1 + \delta E_2$ is the same in both cases.

The change in the energy of migration of different hydrocarbons to water over a given temperature range should therefore be proportional to their coordination number, i.e., be governed by their size. The absolute value of the energy should also depend on the structure of the hydrocarbon, since it depends on the energy difference between the hydrocarbon−water (in the open structure) and hydrocarbon−hydrocarbon interactions.

Figure 2 presents experimental data on the energies of migration of a number of hydrocarbons from a nonpolar environment to water as a function of temperature. It can be seen that the energy of solution becomes more positive as the temperature rises; the slopes of the curves roughly coincide for molecules of similar size (particularly in the case of benzene and butane) and increase with the molecular size. The increase in E_{tr} with rising temperature shows that solution of hydrocarbons in water becomes energetically more disadvantageous as the temperature increases (a minus sign before E_{tr} means that migration of the hydrocarbons to water is energetically advantageous, while a plus sign means that it is energetically disadvantageous). Figure 2 also shows the free energy of migration of hydrocarbons from a nonpolar environment to water as a function of temperature. It can be seen that the free energy also increases with temperature, although more slowly than the energy of migration. The loss of entropy on migration decreases as the temperature rises, so that, while the poor solubility of hydrocarbons in water at room temperature is almost wholly due to the loss of entropy, the migration process becomes quite disadvantageous energetically at high temperatures and the entropy term plays a lesser role.

When compounds containing both polar and nonpolar groups are dissolved in water, the tendency of the nonpolar groups to leave the aqueous environment (the converse process of

Table 2. Free Energy, Energy, and Entropy of Hydrophobic Interaction of Side Chains of Amino Acid Residues and Temperature Functions of Thermodynamic Parameters at 25° (ΔF < 0 indicates attraction of the hydrophobic groups and ΔH > 0 indicates that attraction is energetically disadvantageous)

Amino acid	Paired interactions with side chain of same amino acid					Migration to hydrophobic region		
	ΔF, $\frac{kcal}{mole}$	ΔH, $\frac{kcal}{mole}$	$\Delta S = -\frac{d\Delta F}{dT}$ $\frac{cal}{mole \cdot deg}$	$\frac{d\Delta H}{dT}$ $\frac{cal}{mole \cdot deg}$	$\frac{d\Delta S}{dT}$ $\frac{cal}{mole \cdot}$	ΔF, $\frac{kcal}{mole}$	ΔH, $\frac{kcal}{mole}$	ΔS, $\frac{cal}{mole \cdot deg}$
Alanine . .	—0.7	0.7	4.7	—19.6	—0.066	—1.3	1.5	9.4
Valine . . .	—0.9	1.1	6.7	—29.5	—0.099	—1.9	2.2	13.7
Leucine . .	—0.7	1.1	6.0	—29.5	—0.099	—1.9	2.4	14.3
Isoleucine .	—1.5	1.8	11.1	—49.0	—0.165	—1.9	2.4	14.5
Methionine .	—1.1	1.8	9.8	—49.0	—0.165	—2.0	2.7	16.0
Proline . .	—1.0	1.1	7.1	—29.5	—0.099	—2.0	2.2	14.0
Phenylalanine	—1.4	0.8	7.5	—44.1	—0.148	—0.3	2.7	10.1
						—1.8*	1.0*	9.5*
Cysteine . .	—0.6	0.7	4.4	—19.6	—0.066	—1.6	1.8	11.4
Glutamic acid, glutamine† .	—0.5	0.7	4.0	—	—	—	—	—
Arginine† .	—0.7	1.1	6.0	—	—	—	—	—
Lysine† . .	—1.0	1.4	8.1	—	—	—	—	—
Tyrosine‡ .	—1.2	0.7	6.1	—	—	—	—	—

*The first figure is for migration to an aliphatic environment and the second for migration to an aromatic environment.
†Interaction of nonpolar segments of side chains of polar amino acid residues with leucine.
‡Interaction with phenylalanine.

migration from a nonpolar environment to water) is manifested in their mutual attraction; the free energy of this attraction (hydrophobic interaction) is directly related to the factor considered above, which can be determined experimentally. There is a slight discrepancy resulting from the fact that, when hydrophobic groups form contacts in water, they may not be completely removed from the aqueous environment; the free energy of the hydrophobic interaction of nonpolar groups in molecules containing both polar and nonpolar groups can therefore have a lower absolute magnitude than the free energy of migration of small hydrocarbon molecules, which serve as models of nonpolar groups, from water to a nonpolar environment.

Hydrophobic attraction in the room-temperature region has an essentially entropic character: the nonpolar groups tend to go out of contact with the water, in order not to cause additional structuring. The hydrophobic interaction becomes stronger with temperature as a result of the increase in the contribution made by the energy. Since the energy and entropy of migration of hydrocarbons from water to a nonpolar environment and thus the energy and entropy of the hydrophobic interaction depend on temperature, the free energy of the hydrophobic interaction is a nonlinear function of temperature and, in first approximation, can be represented as a quadratic function of this factor. Detailed theoretical calculations are given in articles by Nemethy and Scheraga and by Silberberg et al. [4, 23]. Table 2 shows the thermodynamic characteristics of the hydrophobic attraction of the nonpolar groups of amino acids in water, as calculated by Silberberg et al. in [23]. It can be seen that the nonpolar groups exhibit strong attraction, which specifically leads to structural compactness of globular proteins. One low-molecular

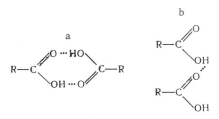

Fig. 3. Cylic (a) and linear (b) carboxylic acid dimers.

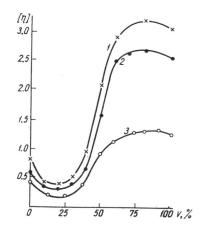

Fig. 4. Characteristic viscosity of uncharged polymethacrylic acid in mixture of water and methanol [27] (v is the methanol concentration by volume). Molecular weight of polymer: 1) $1.0 \cdot 10^{-6}$; 2) $7.1 \cdot 10^5$; 3) $2.7 \cdot 10^5$.

model system in which hydrophobic interactions appear is the series of carboxylic acids [24], $R-COOH$, where R is an aliphatic or aromatic radical $-(CH_2)_nH$, with n = 0, 1, 2, 3, or $-CH_2C_6H_5$. These acids form cyclic dimers in the gaseous phase and in nonaqueous solvents (Fig. 3); the dimerization constant, which is governed by the free energy of dimerization, is virtually independent of the size of the nonpolar R-group. The increase in chain length in aqueous solutions leads to an increase in the tendency to form hydrogen bonds. The free energies of dimerization in the series R = $-CH_3$, $-CH_2CH_3$, $-CH_2CH_2CH_3$, and $-CH_2C_6H_5$ are, respectively, 0.8, 1.0, 1.4, and 1.6 kcal/mole less than the free energy of dimerization of formic acid HCOOH. This is due to formation of noncyclic dimers in water (Fig. 3), where there are contacts between nonpolar groups as well as hydrogen bonds [24]. Although conditions for hydrogen bonding are worse in this case than when cyclic dimers are formed, the corresponding loss of free energy is compensated for by a gain in the energy of the hydrophobic interaction of the nonpolar groups. It should be noted that the free energy of dimerization of formic acid in water is about 2 kcal/mole, i.e., the degree of hydrogen bonding of the carboxyl groups in the absence of hydrophobic interactions is extremely low (about 3%).

Hydrophobic interactions also occur in a model synthetic polymer, polymethacrylic acid $(-CH_2-CCH_3COOH-)_n$, which contains polar and nonpolar methyl groups in each monomeric unit. In an acidified aqueous medium, where the carboxyl groups are not charged, molecules of polymethacrylic acid have a compact structure, occupying a smaller volume in solution than the structurally similar molecules of polyacrylic acid $(-CH_2-CHCOOH-)_n$, which do not contain methyl groups [25-29]. This structure breaks down cooperatively during ionization of the molecules, when the excess of the energy of electrostatic repulsion of the charged groups in the compact structure over the energy of their interaction in the open globular structure normal for synthetic polymers exceeds the energy of hydrophobic attraction of the nonpolar groups. Cooperative breakdown of the compact structure also occurs when methanol, ethanol, or higher alcohols are added to the water, which leads to gradual weakening of the hydrophobic interactions. By way of illustration, Fig. 4 shows the characteristic viscosity of polymethacrylic acid (which is proportional to the volume occupied by the molecule) as a function of solvent composition in a mixture of water and methanol. The abrupt rise in viscosity corresponds to breakdown of the compact structure. It should be noted that calculation of the potential energy of internal rotation in polymethacrylic acid molecules [30] has shown that the short-range ordering in the syndiotactic macromolecule corresponds to a structure favoring formation of contacts between the methyl groups of adjacent segments of the chain.

Hydrophobic interactions of nonpolar molecules and groups in an aqueous medium are thus not characteristic merely of biological macromolecules. They are, however, most pronounced in such molecules, which contain both polar groups responsible for their solubility in water and nonpolar groups whose contact with water leads to an increase in its free energy.

LITERATURE CITED

1. F. M. Richards, Annual Rev. Biochem., 32:369 (1963).
2. S. J. Singer, Advances in Protein Chem., 17:1 (1962).
3. W. Kauzmann, Advances in Protein Chem., 14:1 (1959).
4. G. Nemethy and H. A. Scheraga, J. Chem. Phys., 36:3401 (1962).
5. J. D. Bernal and R. H. Fowler, J. Chem. Phys., 1:515 (1933).
6. H. S. Frank and W. Y. Wen, Disc. Faraday Soc., 24:133 (1957).
7. H. S. Frank, Proc. Roy. Soc. London, A247:481 (1958).
8. O. Ya. Samoilov, Zh. Fiz. Khim., 20:1411 (1946).
9. L. Pauling, The Nature of the Chemical Bond, New York (1960), p. 472.
10. H. S. Frank and A. S. Quist, J. Chem. Phys., 34(2):604 (1961).
11. I. G. Mikhailov and Yu. P. Syrnikov, Zh. Strukt. Khim., 1:12 (1960).
12. G. Nemethy and H. A. Scheraga, J. Chem. Phys., 36:3382 (1962).
13. J. Morgan and B. E. Warren, J. Chem. Phys., 6:666 (1938).
14. M. D. Danford and H. A. Levy, J. Am. Chem. Soc., 84:3965 (1962).
15. W. A. Senior and V. Vand, J. Chem. Phys., 43:1873 (1965).
16. H. S. Frank and M. W. Evans, J. Chem. Phys., 13:507 (1945).
17. T. M. Birshtein and O. B. Ptitsyn, Conformations of Macromolecules [in Russian], Moscow (1964), p. 313.
18. L. I. Dana, A. C. Jenkins, J. N. Burdick, and R. C. Timm, Refrig. Engng., 12:387 (1926).
19. W. F. Claussen and M. F. Poglase, J. Am. Chem. Soc., 74:4817 (1952).
20. T. J. Morrison and F. Billet, J. Chem. Soc., p. 3819 (1952).
21. R. L. Bohon and W. F. Claussen, J. Am. Chem. Soc., 73:1571 (1951).
22. D. M. Alexander, J. Phys. Chem., 63:1021 (1959).
23. G. Nemethy and H. A. Scheraga, J. Phys. Chem., 66:1773 (1962).
24. E. E. Schrier, M. Pottle, and H. A. Scheraga, J. Am. Chem. Soc., 86:3444 (1964).
25. A. Silberberg, J. Eliassaf, and A. Katchalsky, J. Polymer. Sci., 23:259 (1957).
26. V. N. Tsvetkov, S. Ya. Lyubina, and K. L. Bolevskii, in: High-Molecular Carbon-Chain Compounds [in Russian] (1963), p. 26.
27. E. V. Anufrieva, T. M. Birshtein, T. N. Nekrasova, O. B. Ptitsyn, and T. V. Sheveleva, J. Polymer Sci. (In press).
28. J. C. Leite and M. Mandel, J. Polymer Sci., A2:1879 (1964).
29. A. M. Liquori, G. Barone, V. Crescenzi, F. Quadrifoglio, and V. Vitagliano, International Symposium on Macromolecular Chemistry, Prague (1965), p. 588.
30. F. P. Grigor'eva, T. M. Birshtein, and Yu. Ya. Gotlib, Vysokomolek. Soed., A9:580 (1967).

GENERAL PROBLEMS IN THE THEORY OF IONIC HYDRATION IN AQUEOUS SOLUTIONS

O. Ya. Samoilov

N. S. Kurnakov Institute of General and Inorganic Chemistry
Academy of Sciences of the USSR

The present article considers general problems in the theory of ionic hydration in aqueous solutions which must always be kept in mind in studying electrolyte solutions and the processes that take place in them, particularly in biological systems.

We will also attempt to resolve the question of why the course of these processes is to a large extent determined by the state of water, particularly its structure (the short-range ordering of molecular disposition).

Role of Short-Range Forces

In investigating ionic hydration in aqueous solutions, one is struck principally by the fact that an aqueous electrolyte solution is a typical heterodynamic system, i.e., a system in which forces of different types act between particles and a system characterized by interactions of totally different natures. On the one hand, there are interactions among the ions, which slowly weaken with distance and are long-range forces. On the other hand, the interactions between the ions and the water molecules, which are for the most part ion—dipole in character, weaken somewhat more rapidly with distance.

Finally, there are interactions among the water molecules which form hydrogen bonds. These interactions decrease rapidly as the interparticle distance increases and constitute a typical example of short-range forces.

Comparison of the total interaction energies of the particles in aqueous electrolyte solutions, i.e., the energies necessary to remove the interacting particles to an infinitely great distance, yields a picture of ionic hydration in solution in which the state of the system is always governed by the interaction between the water molecules and the ions, the liquid water (solvent) merely serving as a "base" for the molecules that hydrate the ions and are stably bound by them. Actually, the total energy of the interaction of the water molecules is about 4.5–5 kcal per mole, while that of the interaction of the water molecules with the ions is always tens and even hundreds of kilocalories per mole.

However, it has been found that "strong" binding of the water molecules by the ions does not always occur. Such bonding is lacking for cations of the alkali and alkali—earth elements and for the anions Cl^-, Br^-, I^-, NO_3^-, ClO_4^-, etc. Moreover, the disruption of the water structure by the ions, which is not always compensated for by bond formation between the water molecules and the ions, is of great importance for dilute electrolyte solutions containing ions of this type.

One theory to explain the pattern of ionic hydration in aqueous solutions (and of ionic solvation in liquid solutions in general) consists in the following. In liquid solutions (which are condensed systems), an increase in the distance between particles in direct contact at a given instant does not immediately go to physically infinite distances, at which the particle interaction would have no effect, but takes place gradually, at first over short distances, i.e., a certain characteristic length of the order of the particle diameter. This characteristic length is 0.8-1.0 Å in aqueous electrolyte solutions. It must be recognized that the relative particle displacements and hence the binding of particles of, say, type j by particles of type i (e.g., binding of solution water molecules by ions) are governed solely by the changes in interaction energy over short distances near the particles.

Here we can make use of the concepts of the thermal movement of particles in liquids developed by Frenkel' [1]. According to his hypotheses, the particles of the liquid oscillate for some time around temporary equilibrium positions and then, overcoming a potential barrier, are displaced to the next equilibrium potential.

The significance of given interactions in the behavior of particles in liquid solutions is governed by the contribution made by the interaction in question to the magnitude of the potential barrier that the particle must overcome during displacement from a temporary equilibrium position in the solution structure. In general, this contribution is independent of the total interaction energies and depends only on how the interaction energies decrease with distance. The more sharply these energies fall with increasing interparticle distance, the greater is the potential barrier. Interparticle interactions extending over great distances are therefore not as important for solution properties as interactions that are weak but decrease rapidly with distance (short-range forces) [2, 3]. The great importance of short-range interparticle forces for the properties of electrolyte solutions was emphasized by Debye in 1962 [4], in an introductory paper read to a conference on electrolyte solutions.

It is understandable from the foregoing considerations that short-range hydration, i.e., the interaction of ions with the water molecules closest to them, is of special significance in ionic hydration in aqueous solutions. We will henceforth consider short-range hydration.

A General Approach to Investigation of

Ionic Hydration in Aqueous Solutions

Since the state of an ion in solution is governed not by the total interaction energies but principally by the change in energy over short distances near the ion (the "local interaction—energy gradients"), we cannot assert *a priori* that binding of the closest water molecules by the ions occurs in all cases or proceed from the theory of bonding in investigating ionic hydration in solutions. It was necessary to develop a more general approach to the study of ionic hydration in aqueous solutions. I proposed such an approach ten years ago [2, 3, 5]. It consists in considering the exchange of the water molecules closest to the ions for other solution water molecules. Intensification of ionic hydration in the solution corresponds to a decrease in exchange frequency and an increase in the average time (τ_i) for which water molecules in the immediate vicinity of the ion remain in this position.

Let the activation energy of nearest-neighbor exchange in pure water be E, kcal/mole and the average time for which a molecule remains in one equilibrium position be τ, sec. A solution water molecule requires an energy E_i in order to leave the immediate vicinity of the ion, this quantity differing from E in the general case. Let

$$E_i = E + \Delta E_i.$$

Table 1

Ion	ΔE_i	r_i	Ion	ΔE_i	r_i
Li+	0.39	0.68	Cs+	—0.34	1.65
Na+	0.17	0.98	Cl−	—0.10	1.81
K+	—0.20	1.33	Br−	—0.14	1.96
Rb+	—0.30	1.49	I −	—0.15	2.20

The quantities τ_i/τ and ΔE_i can serve as quite general characteristics of ionic hydration in solution. They are roughly correlated by the relationship

$$\frac{\tau_i}{\tau} \cong e^{\frac{\Delta E_i}{RT}}.$$

The approximate character of this relationship is due principally to the fact that the vibration frequency of the water molecules near the ions as a whole is assumed to be the same as in pure water. This assumption is valid at least at low values of ΔE_i. The translational-vibration frequencies of water molecules in aqueous electrolyte solutions and the relationship between these frequencies and the value of ΔE_i have been specifically considered in a previous work [6].

A procedure was devised for estimating the value of ΔE_i for individual ions from the temperature function of ion mobility in solution and the self-diffusion coefficients in water. It was found that there are two cases:

$$1. \; \Delta E_i > 0 \text{ and } \frac{\tau_i}{\tau} > 1 \quad (Mg^{2+}, Li^+, Na^+).$$

$$2. \; \Delta E_i < 0 \text{ and } \frac{\tau_i}{\tau} < 1 \quad (K^+, Cs^+, Cl^-).$$

In the first case, exchange of the water molecules nearest the ions occurs less frequently than exchange of the nearest neighbors in the water, so that we can speak of some effective binding of the nearest solution water molecules by the ions. This case can be termed positive hydration. In the second case, exchange of the water molecules nearest the ions occurs more frequently than exchange of the nearest neighbors in the water. This phenomenon can be called negative hydration. The terms positive and negative hydration are associated with the sign of ΔE_i. The case where $\Delta E_i = 0$ and $\tau_i/\tau = 1$ corresponds to the boundary between positive and negative hydration.

The magnitude of ΔE_i depends on the characteristics of the ion (radius, charge, and electron-shell structure). The values of ΔE_i at 25° for certain single-charge ions, found from the most precise current experimental values for the temperature coefficients of ion mobility in solution [7], are given below. The activation energy of self-diffusion for water E is assumed to be 4.28 kcal/mole (corresponding to the temperature range 15–35°) [8]. Table 1 also shows the crystal-chemical ionic radii r_i in angstroms.

Figure 1 shows ΔE_i as a function of the crystal-chemical ionic radius. The values of ΔE are more precise than those previously given [2]. It must be noted that this increase in accuracy had no effect on the ionic radius corresponding to the boundary between positive and negative hydration. As before, the transition occurs at $r_i^0 = 1.1$ Å.

These results (the high exchange frequency for the water molecules nearest the ions, the existence of negative hydration and of a boundary between positive and negative hydration, etc.) have now received rather broad experimental confirmation in investigations of solutions by various methods.

The study of the change in the entropy of water during ionic hydration made by Krestov [9–11] is of great interest. In essence, this author determined the difference between the entropy of the water molecules closest to the ions and that of the molecules in pure water (or at least established the sign of this difference). He found that the entropy of the water in the

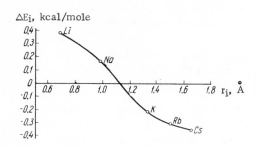

Fig. 1. Dependence of ΔE_i on ionic radius.

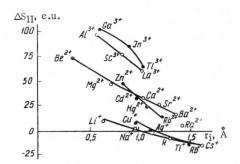

Fig. 2. Dependence of ΔS_{II} on ionic radius for certain cations bearing different charges [9].

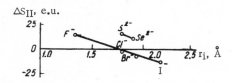

Fig. 3. Dependence of ΔS_{II} on ionic radius for certain anions bearing different charges [9].

immediate vicinity of a number of ions was lower than that of pure water. Moreover, the entropy of the water molecules in the immediate vicinity of certain single-charge ions was higher than that of pure water (this obviously corresponds to negative hydration). In Figs. 2 and 3, which are taken from Krestov's articles, the value of $-\Delta S_{II}$ (the sign of this factor is the reverse of that of the difference in the entropies of the water molecules in the immediate vicinity of the ions and of pure water) is plotted along the ordinate and the crystal-chemical ionic radii are plotted along the abscissa. The boundary between positive and negative hydration found from Krestov's data is in good agreement with that given above: the cations Li^+ and Na^+ are positively hydrated, while K^+, Rb^+, and Cs^+ are negatively hydrated.

Radiospectroscopic methods have yielded valuable results in research on ionic hydration in solutions. We should first mention articles by Valiev [12, 13], who established that the anions I^- and Br^- are negatively hydrated, as well as the work of Shcherbakov [14]. Mazitov [15], who investigated the temperature function of the nuclear magnetic relaxation time of protons and deuterons in aqueous solutions containing the cation Mn^{2+}, established that there is rapid exchange of the water molecules nearest this cation. The activation energy found in this case was close to that calculated from the formula we had previously derived [16], assuming that the boundary between positive and negative hydration corresponds to a single-charge cation with an ionic radius of 1.1 Å. In a study of deuteron relaxation in aqueous solutions of diamagnetic salts, Ionov and Mazitov [17] established that the rotary movement of water molecules is retarded under the action of neighboring Al^{3+}, Mg^{2+}, Li^+, and Na^+ cations and accelerated under the action of K^+ and Cs^+ [17]. This fact indicates the existence of positive and negative hydration.

A very interesting article has been published by Karyakin, Petrov, Gerlit, and Zubrilina [18], who investigated the effect of ions in aqueous solutions on the overtone region (7300-5000 cm^{-1}) of the infrared absorption spectrum of water. They showed that ions fall into two groups. Some ions intensify absorption in the frequency region below the maximum absorption-bond frequency for pure water and somewhat reduce absorption in the frequency region below the maximum. The ions in this group are arranged in the following series in accordance with their effect on the absorption band: $Al^{3+} > Cr^{3+}$, $Be^{2+} > Cd^{2+} > Zn^{2+} > Mg^{2+}$, $Li^+ > Na^+$ (Na^+ has virtually no effect), $CO_3^{2-} > SO_4^{2-}$, and $OH^- > F^-$. Ions of the other group reduce the absorption in the low-frequency region and somewhat intensify it in the region above the maximum. The ions in this group are arranged in the series $Cs^+ > K^+$ and $ReO_4^- > ClO_4^- > I^- > NO_3^- > SCN^- >$

Cl^-. The ions in the first group strengthen the bonds between the solution water molecules and the nearest neighbors, while the ions in the second group weaken these bonds. The validity of separating ions into two groups, the first including Al^{3+}, Mg^{2+}, Li^+, and Na^+ and the second including K^+ and Cs^+, is underscored by the fact that two simultaneous effects are observed in the presence of an anion belonging to the second group and a cation belonging to the first group. For example, this phenomenon occurs in the case of aluminum perchlorate.

The differing hydration of Li^+ and Na^+ on the one hand and K^+ and Cs^+ on the other can also be seen from the results of dielectric measurements made by Yastremskii [19, 20] (the dielectric permeability and dielectric-loss coefficient of aqueous electrolyte solutions at a frequency of 9400 mHz).

These hypotheses regarding the short-range hydration of ions in aqueous solutions are in agreement with the results of studies of the temperature function of the coordination numbers of ions in dilute aqueous solutions (the average numbers of water molecules composing the coordination shell closest to the ion). It has been demonstrated [21] that

$$\frac{\partial}{\partial T}\left(\frac{n_i}{n}\right) < 0, \text{ when } \Delta E_i > 0,$$

and

$$\frac{\partial}{\partial T}\left(\frac{n_i}{n}\right) > 0, \text{ when } \Delta E_i < 0,$$

where n_i is the coordination number of the ion in solution and n is the coordination number of the molecules in the water. The temperature function of n is known from x-ray data [2]. Experimental investigation of the temperature function of the coordination numbers n_i of alkali-metal cations and halide anions in dilute aqueous solutions by the thermochemical method [22] showed that $(\partial/\partial T)(n_i/n) < 0$ for the Li^+ cation and $(\partial/\partial T)(n_i/n) > 0$ for the K^+, Rb^+, and Cs^+ cations and the Br^- and I^- anions. The value of n_i for Cl^- and Na^+ varies with temperature in roughly the same fashion as n, so that $(\partial/\partial T)(n_i/n) \approx 0$; it must, however, be noted that the temperature function of the coordination number of K^+ should be determined by an independent method [21].

It thus turns out that the short-range hydration of Na^+ and K^+ differs in sign: Na^+ is characterized by positive hydration, while the hydration of K^+ is negative. This result (the differing hydration of Na^+ and K^+) is apparently very important for interpreting many of the processes that occur in biological systems.

It should be noted that negative hydration occurs when the contribution of the interaction of the ion with the nearest water molecule to the activation energy of translational movement of this molecule is less than the corresponding contribution of the nearest-neighbor interaction in the water. The presence of "free" water in the solution is not a necessary condition for negative hydration. The fact that the proposed approach to investigation of hydration pertains to dilute solutions results from consideration of the interaction of only one ion with the nearest neighboring water molecule. Extension of this approach to concentrated solutions requires that the influence of the other ions be taken into account. These problems have now been considered in articles on the theory of the salting-out of electrolytes from aqueous solutions [23].

Negative solvation occurs only in those cases where the interaction of the solvent molecules is sufficiently strong (the relative molecular ordering in such solvents is substantial). It must also be noted that each molecule in the water participates in an average of four hydrogen bonds [24]. Krestov showed that, even in such liquids as methanol and ethanol, negative solvation is not observed at a temperature of 25° [25]. Negative hydration is succeeded by positive hydration in aqueous solutions as the temperature rises: the boundary between positive

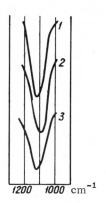

Fig. 4. Absorption bands of Cl—O bond. 1) In anhydrous sodium perchlorate; 2) in crystalline hydrate; 3) in aqueous sodium perchlorate solution (the ordinates correspond to percentage transmission).

and negative hydration is displaced toward larger crystal-chemical ionic radii [26] (the exchange frequency for the molecules near the ions at elevated temperatures is comparable to that for the molecules in the water, whose structure is greatly disrupted by the thermal movement of the particles). The high exchange frequency for the water molecules near the ions in aqueous solutions produces a very large exchange value for ions in solution. Investigation of the exchange of the water molecules closest to ions is therefore not some narrow problem in the study of solutions, but is of very great general importance, since determination of the state of ions in solution is associated with it.

Role of Solvent State

The interaction of water molecules and the structure of water are governed by the hydrogen bonds formed among the molecules and are hence regulated by short-range forces [24]. It is for precisely this reason that the water—water interaction is of great significance for the behavior of particles in aqueous solutions and to a large extent determines their properties. This is especially true of comparatively dilute solutions containing "free" solvent, i.e., solvent whose molecules are not incorporated into the immediate ionic environment.

Formation of bonds among water molecules leads to a decrease in ionic hydration. An interesting illustration of this is provided by the results given in the previously cited article by Karyakin, Petrov, Gerlit, and Zubrilina [18]. These authors investigated the influence of the nearest water molecules on the infrared absorption spectra of the oxygen-containing anions ClO_4^- and SO_4^{2-} in crystalline hydrates of sodium perchlorate and sodium sulfate and in aqueous solutions of these salts. The absorption bands for the Cl—O and S—O bonds in the crystalline hydrates underwent substantial changes under the action of the crystal water. This made it possible to establish that hydrogen bonds are formed between anions and crystal water. Figure 4 (in which the ordinates correspond to percentage transmission) shows the absorption bands for the Cl—O bond in anhydrous sodium perchlorate (curve 1), in its crystalline hydrate (curve 2), and in an aqueous $NaClO_4$ solution (curve 3). The absorption maximum in the anhydrous salt occurs at 1115 cm^{-1}, while that in the crystal hydrate is displaced toward lower frequencies and occurs at 1085 cm^{-1}. This displacement indicates formation of hydrogen bonds between the water molecules and the oxygens of the ClO_4^- anions. The displacement of the band disappears in aqueous solution (curve 3) and the absorption maximum occurs at almost the same frequency as in anhydrous $NaClO_4$. A similar pattern was observed for the sulfates. It was thus found that, as one moves from crystalline hydrates to aqueous solutions, i.e., as the water content of the system increases, the effect of the water on the Cl—O and S—O absorption bands becomes substantially less pronounced. The authors interpret this interesting fact quite correctly. There is undoubtedly a weakening of anionic hydration as a result of formation of water—water bonds.

Conversely, rupture of the water—water bonds leads to an intensification of ionic hydration in solution. Thus, ionic hydration increases to a certain extent as the temperature is raised and the water structure breaks down. This can specifically be seen from the results obtained by Zagorets, Ermakov, and Grunau [27], who measured the spin-lattice relaxation time of protons in solutions of LiCl in water and $CoCl_2$ and $CuCl_2$ in methanol, using different solution

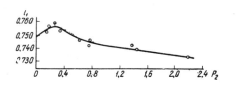

Fig. 5. Solubility of NaCl in methanol
with ethanol added.

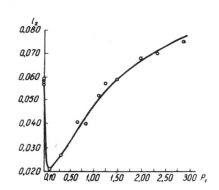

Fig. 6. Solubility of NaCl in ethanol
with methanol added.

concentrations and different temperatures (between −30 and 40°). It was shown in an article on the theory of salting-out from aqueous solutions [28] that hydration of Ca^{2+} in aqueous solution is somewhat intensified as the temperature is raised over the 15–35° range. The foregoing considerations regarding the role of solvent state in ionic solvation in solution naturally does not pertain merely to aqueous solutions.

The results of an investigation of the solubility of sodium chloride in mixtures of ethanol and methanol [29] are interesting in connection with the problem of the role of solvent state in solvation. Addition of small amounts of the second component can have different effects on the structure of the liquid, depending on the ratio of the molecular sizes of the main component and the additive and on the character of the bonds formed. Addition of small amounts of methanol to ethanol leads to stabilization of the ethanol structure. Addition of small amounts of ethanol to methanol causes breakdown of the methanol structure. The molecular mechanism of these structural effects has been investigated. Stabilization of the solvent structure should lead to a decrease in solvation and thus in the solubility of the NaCl. Breakdown of the structure should have the opposite effect. This was the situation observed experimentally (Figs. 5 and 6). The notations l_1 and l_2 on the graphs are the solubilities of NaCl (mol.%) in methanol with ethanol added (P_2 mol.%) and in ethanol with methanol added (P_1 mol.%). The results obtained are especially conclusive because the solubility of NaCl in methanol is higher than that in ethanol. The low additive concentrations at which solubility anomalies occur prove the structural character of the effects observed.

An increase in free-solvent ordering (structural stabilization) thus leads to a decrease in solute solvation. Conversely, breakdown of the free-solvent structure intensifies solvation. The role of solvent state in ionic solvation in dilute solutions underscores the great importance of short-range forces in the properties of liquid solutions.

The change in macromolecular hydration (in the sense of weakening or strengthening of the bonds to the closest water molecules) under the influence of stabilization or breakdown of the free-water structure is undoubtedly very important in biology. The difference in the character of the hydration of Na^+ and K^+ is of special significance.

LITERATURE CITED

1. Ya. I. Frenkel', Kinetic Theory of Liquids [in Russian], Izd. AN SSSR, Moscow–Leningrad (1945).
2. O. Ya. Samoilov, Structure of Aqueous Electrolyte Solutions and the Hydration of Ions, Consultants Bureau, New York (1965).
3. O. Ya. Samoilov, Disc. Faraday Soc., No. 24, 141 (1957).
4. P. Debye, Electrolytes, Pergamon Press, New York (1962).
5. O. Ya. Samoilov, Izv. Akad. Nauk SSSR, Otd. Khim. Nauk, No. 2, p. 242 (1953).
6. O. Ya. Samoilov, Zh. Strukt. Khim., 1(1):36 (1960).
7. R. A. Robinson and R. Stokes, Electrolyte Solutions, Butterworths Scientific Publishers, London (1959).

8. G. A. Andreev, Zh. Fiz. Khim., 37(2):361 (1963).
9. G. A. Krestov, Zh. Strukt. Khim., 3(2):137 (1962).
10. G. A. Krestov, Izv. Vyssh. Uch. Zav., Khim. i Khim. Tekhnol., 8(5):734 (1965).
11. G. A. Krestov, Zh. Strukt. Khim., 3(4):402 (1962).
12. K. A. Valiev, Zh. Strukt. Khim., 3(6):653 (1962).
13. K. A. Valiev, Zh. Strukt. Khim., 5(4):517 (1964).
14. V. A. Shcherbakov, Zh. Strukt. Khim., 2(4):484 (1961).
15. R. K. Mazitov, Dokl. Akad. Nauk SSSR, 152(2):375 (1963).
16. E. S. Lekht and O. Ya. Samoilov, Zh. Strukt. Khim., 3(4):466 (1962).
17. V. I. Ionov and R. K. Mazitov, Zh. Strukt. Khim., 7(2):184 (1966).
18. A. V. Karyakin, A. V. Petrov, Yu. B. Gerlit, and M. E. Zubrilina, Teor. i Éksperim. Khim., 2(4):494 (1966).
19. P. S. Yastremskii, Zh. Strukt. Khim., 2(3):268 (1961).
20. P. S. Yastremskii, Zh. Strukt. Khim., 3(3):279 (1962).
21. O. Ya. Samoilov, Dokl. Akad. Nauk SSSR, 126(2):330 (1959).
22. M. N. Buslaeva and O. Ya. Samoilov, Zh. Strukt. Khim., 2(5):551 (1961).
23. O. Ya. Samoilov, Zh. Strukt. Khim., 7(1):15 (1966).
24. O. Ya. Samoilov and T. A. Nosova, Zh. Strukt. Khim., 6(5):798 (1965).
25. G. A. Krestov, Zh. Strukt. Khim., 3(5):516 (1962).
26. G. A. Krestov and V. K. Abrosimov, Zh. Strukt. Khim., 5(4):510 (1964).
27. P. A. Zagorets, V. I. Ermakov, and A. P. Grunau, Zh. Fiz. Khim., 39(7):1955 (1965).
28. O. Ya. Samoilov, Zh. Strukt. Khim., 7(2):175 (1966).
29. O. Ya. Samoilov, I. B. Rabinovich, and K. T. Dudnikova, Zh. Strukt. Khim., 6(5):768 (1965).

STRUCTURE OF WATER IN CRYSTALLINE HYDRATES OF CERTAIN BIOLOGICALLY IMPORTANT COMPOUNDS

G. G. Malenkov

N. S. Kurnakov Institute of General and Inorganic Chemistry,
Academy of Sciences of the USSR

Determination of the precise localization of water molecules interacting with the functional groups of biopolymers is a very complex task. Our knowledge of the position and state of the water molecules in supermolecular structures (nucleoproteins and lipoproteins) is even less definite. However, it is known that they incorporate water, which serves both a structural and a functional role.

In order to clarify our notions of the crystal-chemical characteristics of water in connection with a study of its possible participation in the formation of biological structures, we became interested in analyzing the experimental data that have been amassed on the structures of crystalline hydrates of compounds whose molecules are related to the components of biopolymers. The position and, in many cases, the orientation of the water molecules in crystals of such low-molecular compounds can now be determined very accurately.

We will consider below the structures of hydrates of amino acids, oligopeptides, purine and pyrimidine bases, nucleosides, nucleotides, and certain other compounds.

The work of Kitaigorodskii's school [1, 2] conclusively demonstrated that the molecular arrangement in organic crystals (and in molecular crystals in general) obeys the principle of densest packing. This principle plays an even larger role in organic crystal chemistry than in, for example, the crystal chemistry of ionic or metallic compounds, since, in view of the characteristics of the intermolecular-interaction forces, there are better grounds for ascribing a definite geometric shape to molecules than to ions. Moreover, particles of complex shape can usually be packed more densely than spherical particles.

That densest packing occurs in crystals formed by nonpolar molecules is more or less obvious. However, the molecules in crystals of compounds containing hydrogen, which is capable of participating in hydrogen bonding, are also very densely packed; the dense-packing requirement does not conflict with the tendency of the molecules to form hydrogen bonds with one another [1].

One of the few examples of a molecular crystal in which the tendency of the molecules to form the maximum number of hydrogen bonds conflicts with the tendency toward densest packing is apparently the ice crystal. This contradiction is not resolved at moderate pressures and the two requirements are satisfied by formation of two interpenetrating hydrogen−bond systems only in the structure of ice VII (which exists at pressures above 22,000 atm) [3].

Incorporation of water molecules into organic crystals, i.e., copacking of the very small water molecules and the larger molecules of the main compound, is to be expected when such copacking leads to better space filling and permits the molecules to be arranged in such a fashion that the conditions for hydrogen bonding are most favorable. The number of water molecules and their state are governed by the shape of the main-compound molecules, by the arrangement of the hydrophilic groups, and by certain specific characteristics of the intermolecular interaction (e.g., the interactions in a stack of parallel cyclic molecules and the electrostatic interactions that occur in the case of ionized molecules or functional groups).

Copacking is often effected in such fashion that the water molecules are located singly among the main-compound molecules. In this case, each H_2O molecule can form one, two, three, or four hydrogen bonds with different lengths. For example, each water molecule in hydrated dialuric acid [4]

forms four hydrogen bonds having lengths of 2.68, 2.74, 2.89, and 2.90 Å with the hydroxyl groups of the acid. The two-dimensional layers of molecules joined by hydrogen bonds thus formed are arrayed with respect to one another in such fashion that the heterocyclic molecules form stacks with a distance of 3.56 Å between the molecular planes.

Hydrated deoxyadenosine [5] can serve as an example of a crystal in which single water molecules form three hydrogen bonds each. Water molecules form hydrogen bonds having lengths of 2.73, 2.65, and 2.78 Å with the hydroxyl groups of deoxyribose (OH−H_2O−OH angles of 118.5, 120.3, and 11.9°); adenine forms hydrogen bonds with adenine on a standard pattern and is packed into stacks with a distance of 3.67 Å between the planes of the heterocyclic rings.

The water molecules of hydrated adenosine phosphate [6] fit very well between the heterocyclic portion of one molecule and the phosphate group of another, forming two comparatively weak hydrogen bonds (2.863 and 2.824 Å at an angle of 123°) on the pattern:

The water molecules in hydrated leucyl−prolyl−glycine [7] form two still weaker hydrogen bonds with the oxygen atoms of the carboxyl groups (2.936 and 2.809 Å) of different molecules.

The role of single water molecules in holding together the hydrophilic groups of molecules and promoting better packing of the hydrophobic groups can be strikingly illustrated with hydrated glycyl-tryptophan (Fig. 1) [8]. As is shown in the figure, this crystalline hydrate contains water molecules of two types, which form two or three hydrogen bonds with the oxygen atoms of the carboxyl groups, the keto oxygen, and the glycyl amino group: H_2O_1−O (2.99), −O (2.77), −N (2.68), H_2O_{II}−N (2.75), and −O (2.74).

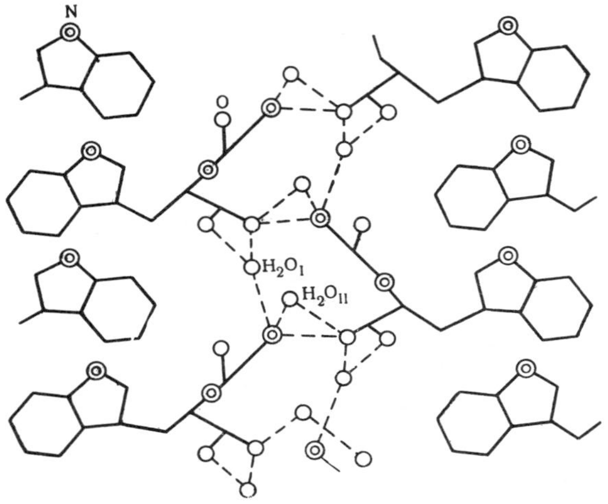

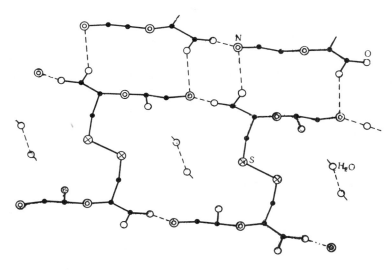

Fig. 1. Diagram of structure of glycyl-l-tryptophan.
The broken lines represent hydrogen bonds.

Fig. 2. Structural projection of hydrated diglycyl-l-cystine.

The main-compound molecules and pairs of hydrogen-bonded water molecules (with a bond length of 2.69 Å) are copacked in the dimethyloxypyrimidine dihydrate crystal [1]. Each water molecule forms one hydrogen bond with a hydroxyl group (2.98 and 2.71 Å) and one with a nitrogen atom (2.75 and 2.90 Å). The dimethyloxypyrimidine molecules do not form hydrogen bonds with one another. This structure is characterized by close approximation of the heterocyclic molecules in the stacks (3.38 Å).

The water molecules in fluoroglucitol dihydrate (see [9]) are also paired, but they form four tetrahedrally directed hydrogen bonds (three to hydroxyl groups). The bonds are oriented in such fashion that the bond between the water molecules is centrosymmetric in character.

In a number of cases, channels develop when organic molecules are arranged into a structure; these can be filled with chains of water molecules. Copacking of such chains and main-

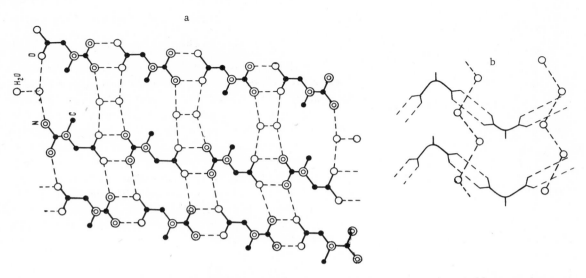

Fig. 3. Projection of structure of creatine hydrate (diagrammatic). a) Along water-molecule chains; b) in (100) plane, water-molecule and creatine-molecule chains visible.

compound molecules leads to very good space filling. If the channel is sufficiently wide, the water molecules may form almost no hydrogen bonds with the organic molecule. Thus, for example, the $N-H_2O$ distance of 3.13 Å in diglycylcystine dihydrate [9, 10] is apparently too great for hydrogen bonding (Fig. 2).

Structures containing chains of water molecules must satisfy one condition associated with the crystal-chemical characteristics of water, i.e., the identity period along the chain should be close to 4.5 Å [9]. This feature may influence the packing of the main-compound molecules. If the organic molecules are linear, they may take on a definite conformation in order to make their packing compatible with the chain of water molecules. For example, the amino acid molecules in creatine hydrate [11] are curved in such a fashion that the identity period along the axis perpendicular to the length of the molecule is 4.5 Å. The creatine molecules in this structure form chains perpendicular to the water-molecule chains (Fig. 3). The water molecules participate in four hydrogen bonds.

The arginine molecules [12] in an arginine dihydrate crystal form chains similar to the creatine-molecule chains. However, the arginine molecules, being longer, provide room for two water-molecule chains per amino acid molecule. On the other hand, the arginine molecule cannot adopt a conformation such that the packing has a spacing of 4.5 Å. The chains consequently break up into pairs (the distance between the molecules of adjacent pairs is 3.3 Å), which draw together and produce squares of water molecules. The water molecules in a square tie together the "head" and "tail" (the guanidine and carboxyl groups) of each arginine molecule, producing the conformation observed in the crystal (Fig. 4).

As has been repeatedly mentioned, cyclic molecules are characterized by a tendency to form stacks. Since the distance between the molecular planes should be of the order of 3.5 Å, the planes of the stacked molecules should be inclined to the translation axis in order for the stack to be compatible with a water chain. Figure 5 shows one way in which this can be effected. The cyclic molecules are joined into pairs by hydrogen bonds. The angle of inclination

G. G. MALENKOV

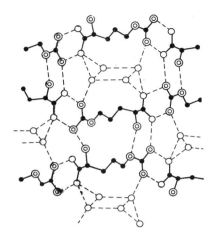

Fig. 4. Projection of structure of *l*-arginine hydrate.

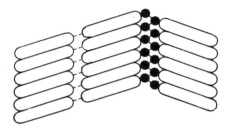

Fig. 5. Copacking of cyclic molecules and water-molecule chains.

of the molecular planes to the translation axis should be $\arcsin\frac{3.5}{4.5} \approx 55°$. An example of a structure in which this situation occurs is theophyllin hydrate [13] (Fig. 6).

A different pattern is observed in thymine hydrate, whose molecules can be joined into flat strips by hydrogen bonds. The molecular planes in this crystal are perpendicular to the axes of the channels through which the water-molecule chains run. The chains are incompatible with the packing of the organic molecules in this case. The best interpretation of this type of structure was given by Gerdil [14], who suggested that the water-molecule chain breaks up, some of the molecules drop out, and the others are displaced from their original positions in such a fashion that the distance separating them reaches the normal value (i.e., 2.7-2.8 Å) (Fig. 7). It is extremely difficult to determine the precise localization of the water molecules in such crystals, but Gerdil's interpretation is quite natural and provides the best explanation of the observed diffraction pattern. Similar breakage of the chains and disordered displacement of the water molecules is observed in caffeine hydrate [15] and in the crystalline hydrate of the noncyclic compound biuret [16].

The chain sometimes breaks "regularly" and the water molecules occupy rigidly fixed positions, separating into pairs (e.g., in arginine hydrate) or triplets (in dihydroxybarbituric acid trihydrate [17]).

The molecules in tetrahydroxybenzoquinone hydrate [18] form strips similar to those in thiamine hydrate, but they are displaced relative to one another in such fashion that the channels capable of holding water molecules run at a large angle to the strip planes and the water-molecule chain may even be somewhat stretched (Fig. 8).

A very interesting structure containing a branched water-molecule chain is the hexahydrate of dilituric acid [19] (Fig. 9).

This crystal is characterized by a very low packing factor, about 0.4 (taking into account the hydrogen atoms on the nitrogen and naturally disregarding those on the water molecules); the acid molecules are joined by hydrogen bonds and form chains similar to those in the thymine crystal. These chains are superimposed on one another in such fashion that stacks of dilituric acid molecules with an interplanar distance of 3.23 Å are formed. The stacked chains are not densely packed and the channels between them are rather wide, running along the stacks and providing room for a complex, branched arrangement of water molecules (Fig. 9). This arrangement, like any rather complex arrangement of tetrahedrally bonded water molecules, is very open. The structure is stabilized by a system of hydrogen bonds and by the intermolecular

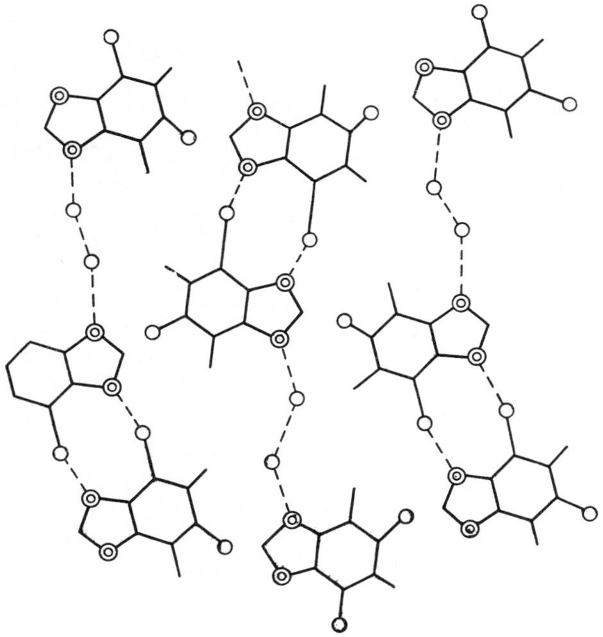

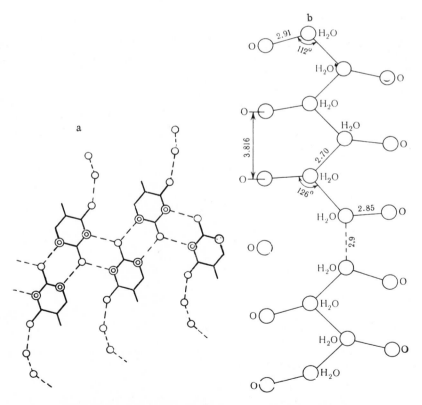

Fig. 6. Diagram of structure of theophyllin hydrate; view along water-molecule chains.

Fig. 7. Structure of thymine hydrate. a) Projection along water-molecule chains; b) broken water-molecule chains.

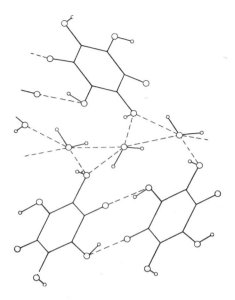

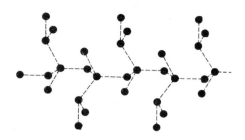

Fig. 9. Arrangement of water molecules in hydrated dilituric acid crystal.

Fig. 8. Water-molecule chains in tetrahydroxybenzoquinone hydrate; hydrogen
atoms shown.

interaction within the stack. In contrast to
a simple flat chain, a spacing of 4.5 Å is not
required for a complex arrangement of water
molecules and the spacing along it equals
twice the thickness of the cyclic molecules.

Let us now turn to hydrate structures
in which electrostatic interactions as well
as hydrogen bonds and intermolecular forces
play a major role. Among the crystals of
low-molecular compounds we have considered, we must deal with electrostatic interactions in,
for example, crystalline halides and phosphates of organic bases.

The $-NH_3^+$ groups, halogen ions, and water molecules in crystalline hydrohalides of organic bases and amino acids form various structures. Chains of alternating cations and anions
separated by water molecules are sometimes observed:

as in the hydrated hydrochlorides of isoleucine [20] and histidine [21].

The cation and anion sometimes form ionic pairs joined by water molecules:

$$....H_2O....I^-....H_2O....I^-....H_2O....I^-....$$
$$NH^+_3 \qquad NH^+_3 \qquad NH^+_3$$

Arrangements of this type are characteristic of the semihydrates of adenine and guanine hydrochlorides [22, 23].

Hydrated thiamine hydrochloride [24] exhibits a continuous cation−anion chain, in which
each chlorine ion is hydrated by a single water molecule:

$$....Cl^-....NH^+_3....Cl^-....NH^+_3....Cl^-.. .NH^+_3....$$
$$H_2O \qquad H_2O \qquad H_2O$$

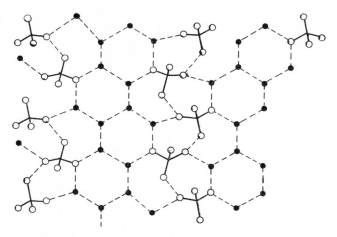

Fig. 10. Anionic layer in crystal of spermine phosphate hydrate.

Table 1. Distances Separating NH_3^+, Cl, and H_2O in a Number
of Crystals

Compound	$dH_2O \ldots Cl^-$, Å	$dH_2O \ldots N^+$, Å	$dCl^- \ldots N^+$, Å
Isoleucine	3.24	2.96	—
Histidine	3.194	2.817	—
Guanine	3.12; 3.16	—	3.18
Adenine	3.12	—	3.18
Thiamine	3.163	—	3.24

Table 1 shows the distances separating the NH_3^+, Cl, and H_2O in the aforementioned crystals.

Water is far more important in crystals of organic phosphate and hydrophosphates. In the structures of thiamine pyrophosphate trihydrate and spermine phosphate hexahydrate, the $H_2PO_4^-$ ions and water molecules form loose anionic layers in which hydrogen bonds play an important structural role; these alternate with more densely packed cationic layers.

In the thiamine pyrophosphate crystal [25], the $H_2PO_4^-$ anions split off from the molecules of the compound

$$\left[H_3C \overset{N}{\diagdown} \overset{NH_2}{\diagdown} \quad \overset{S}{\diagdown} \quad CH_2CH_2O-\overset{\overset{O}{\|}}{\underset{O}{P}}-OH \right]^+ \cdot [H_2PO_4]^-$$

The water molecules in the anionic layer are joined by hydrogen bonds, forming rings with the "chair" configuration so characteristic of the ice structure:

H_2O_I ---- H_2O_{II} ---- H_2O_{II}

H_2O_{II} ---- H_2O_{II} ---- H_2O_I

$H_2O_I — H_2O_{II} — 2,908$ Å

$H_2O_{II} — H_2O_{II} — 2.789$ Å

$< H_2O_{II} — H_2O_I — H_2O_{II} — 87°$

$< H_2O_{II} — H_2O_{II} — H_2O_I — 100°$

Water molecules can be hydrogen bonded to dihydrophosphate anions, forming dimers of the type

$$\begin{array}{ccccccc} OH & & O & & HO & & O \\ & \diagdown & \diagup & & & \diagdown & \diagup \\ & P & & \cdots\cdots & & P & \\ & \diagup & \diagdown & & & \diagup & \diagdown \\ OH & & O & \cdots\cdots & HO & & O \end{array}$$

The hydrogen bond $=O\cdots H-O$ is very strong (2.484 Å long). The phosphates in the cationic and anionic layers also form hydrogen bonds with one another.

The anionic layer in spermine phosphate hexahydrate is even more intricately constructed (Fig. 10).

The spermine cations $\overset{+}{N}H_3 - (CH_2)_3 - \overset{+}{N}H_2 - (CH_2)_3 - \overset{+}{N}H_2 - (CH_2)_3 - NH_3$ are packed between the anionic layers. The nitrogen atoms form weak hydrogen bonds with the oxygen atoms of the phosphate groups.

Unfortunately, too little research has been done on the structure of crystalline hydrates of low-molecular organic compounds of definite biological significance, particularly those that participate in the formation of biopolymers. In this brief survey, we have discussed virtually all structures of this type that have been studied. We have purposely not touched on the structure of crystals containing inorganic cations, although a number of conclusions drawn from analysis of their crystal chemistry could undoubtedly aid in interpreting the biophysics and physics of polymers. This should, however, form the subject of a special review.

Despite the limited data available, we can apparently still use specific structured materials to trace the influence of the packing of the organic molecules on the arrangement of the water molecules and, conversely, the influence of the water molecules on the packing and conformation of the organic molecules. In addition, we can elucidate a number of the other crystal-chemical characteristics of water-containing organic crystals. The author hopes that the data in this survey will be useful to persons interested in the interaction of biopolymers with water.

LITERATURE CITED

1. A. I. Kitaigorodskii, Organic Chemical Crystallography, Consultants Bureau, New York (1961).
2. A. I. Kitaigorodskii, Acta Crystallogr., 18:585 (1965).
3. B. Kamb and B. Davis, Proc. Nat. Acad. Sci. USA, 52:1433 (1954).
4. L. E. Alexander and D. T. Pitman, Acta Crystallogr., 9:501 (1956).
5. D. Watson and D. F. Sutor, Acta Crystallogr., 19:111 (1965).
6. J. Kraut and L. H. Jensen, Acta Crystallogr., 16:79 (1963).
7. Y. C. Leung and R. E. March, Acta Crystallogr., 11:17 (1958).
8. R. A. Pasternak, Acta Crystallogr., 9:341 (1956).
9. G. G. Malenkov, Zh. Strukt. Khim., 3:220 (1962).
10. H. L. Yakel and E. Hughes, Acta Crystallogr., 7:291 (1954).
11. H. Mendel and D. Hodgkin, Acta Crystallogr., 7:443 (1954).
12. I. L. Karle and I. Karle, Acta Crystallogr., 17:835 (1964).
13. D. F. Sutor, Acta Crystallogr., 11:83 (1958).
14. V. A. Shibnev, V. N. Rogulenkova, and N. S. Andreeva, Biofizika
15. D. F. Sutor, Acta Crystallogr., 11:453 (1958).
16. E. Hughes and H. L. Yakel, Acta Crystallogr., 14:345 (1961).
17. D. Mootz and G. A. Jeffrey, Acta Crystallogr., 19:717 (1965).
18. H. P. Klug, Acta Crystallogr., 19:983 (1965).
19. B. M. Craven, S. Martinez-Carrera, and G. A. Jeffrey, Acta Crystallogr., 17:897 (1964).

20.   J. Trommel and J. M. Bijvoet, Acta Crystallogr., 7:703 (1955).
21.   F. Donohue, L. K. Lavine, and F. S. Kolett, Acta Crystallogr., 9:65 (1956).
22.   W. Cochran, Acta Crystallogr., 4:81 (1952).
23.   F. M. Broomhead, Acta Crystallogr., 4:92 (1951).
24.   F. Braut and H. Reed, Acta Crystallogr., 15:747 (1962).
25.   F. L. Karle and K. Britts, Acta Crystallogr., 20:118 (1966).
26.   Y. Iitaka and J. Huse, Acta Crystallogr., 18:110 (1965).

MANIFESTATION OF WATER STRUCTURE IN THERMAL DENATURATION OF MACROMOLECULES

P. L. Privalov

Institute of Proteins
Academy of Sciences of the USSR

It is well known that water, to a considerable extent, governs macromolecular configuration. Nevertheless, one often loses sight of the fact that this property of water is directly due to its structure, which can in turn be altered under the action of various factors. In particular, the structure of water changes with temperature and its influence on macromolecules should thus also vary with temperature. In considering thermal denaturation, however, the temperature is generally assumed to be a factor that acts only on the macromolecule and the problems of the water structure and the change in the macromolecule−water interaction are quietly bypassed. As a result of this ignoring of the solvent, we know absolutely nothing about many of the phenomena associated with the conformational transformations of macromolecules in water, which are apparently of very general significance.

As examples of such phenomena that cannot be interpreted without a more detailed consideration of the properties of water, we can cite the following two facts, which we had previously noted and which evoked only astonishment for a long time.

The first fact is that there is a pronounced increase (by about 20%) in the heat capacity of macromolecules during thermal denaturation. This can be clearly seen in Fig. 1, which shows the partial heat capacity of egg albumin as a function of temperature. The peak corresponds to absorption of heat during denaturation.

The increase in heat capacity in the denatured state cannot be attributed to transition from a rigid compact structure to a chaotic labile cluster having a substantially larger number of degrees of freedom: the increment in heat capacity exceeds the figure calculated for this case.

The second fact is that the enthalpy of macromolecular denaturation is highly dependent on temperature.

As is well known, the principal condition for a cooperative intramolecular transformation of the fusion type is that the free energy equals zero:

$$\Delta E_m = \Delta H_m - T_m \Delta S_m = 0.$$

Hence, it follows that $\Delta H_m = T_m \Delta S_m$ during denaturation. It is generally presumed that the value of ΔS_m (the denaturational increment in entropy) does not depend to any large extent on ambient conditions.

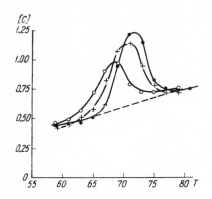

Fig. 1. Partial heat capacity of egg albumin as a function of temperature.

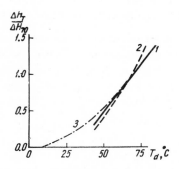

Fig. 2. Relative change in enthalpy of denaturation of DNA (1), egg albumin (2), and chymotrypsin (3) during variation of denaturation temperature.

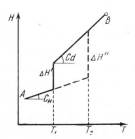

Fig. 3. Enthalpy of macromolecular solution as a function of temperature.

Assuming that ΔS_m is constant in the first approximation, as is usually done in current denaturation theories, we find that the enthalpy of the transformation must be a linear function of the absolute temperature.

However, it actually turns out that this dependence is substantially more abrupt.

This can be seen from Fig. 2, which shows the temperature functions of the enthalpy of denaturation obtained for egg albumin and DNA in our laboratory by direct microcalorimetric measurement [1, 2] and for chymotrypsinogen by Brandt [3], who proceeded from equilibrium data. For ease of comparison, the graph gives the change in enthalpy with respect to its value at 70°.

The most interesting fact is that the curves have almost the same slope in all three cases, although they are totally different. This undoubtedly indicates that we are dealing with a rather general regularity; it should therefore have the same cause in all cases.

It can easily be demonstrated that the two facts discussed above are different aspects of the same phenomenon. Let us consider the graph for the enthalpy of a macromolecular solution as a function of temperature (Fig. 3). Taking into account the fact that the slope of the curve is directly determined by the heat capacity (i.e., dH/dT = C), it is clear that the slope should increase after denaturation. Let us assume that denaturation can occur at the temperatures T_1 and T_2. As can be seen from the graph, the enthalpy difference should be greater in the second case, i.e., the enthalpy of transition should be temperature-dependent.

It thus becomes obvious that we are dealing with a single very general phenomenon, which is little understood from the standpoint of the usual conformational statistics of polymers. However, the picture becomes considerably clearer if we turn our attention to the structural characteristics of water and the mechanism of the water—macromolecule interaction.

Let us consider the factors that stabilize macromolecular conformation.

Hydrophobic and hydrogen interactions are receiving most attention in this respect at present.

Hydrophobic interactions are known to result from the capacity of water to undergo ordering near nonpolar groups and from the undesirability of such ordering, since it is entropically disadvantageous. The nonpolar groups therefore develop a tendency to aggregate, in order to reduce the number of contacts with the water and thus the ordering effect on it. Hence it follows that formation of a hydrophobic bond must be accompanied by absorption rather

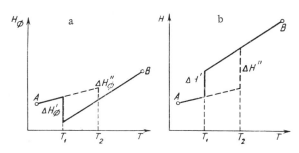

Fig. 4. Temperature function of enthalpy of system containing: a) only hydrophobic bonds that rupture cooperatively at a temperature T_1 or T_2; b) hydrophobic and hydrogen bonds.

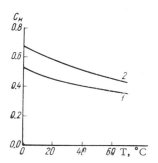

Fig. 5. Concentrations of hydrogen bonds in pure water (1) and in presence of aliphatic alcohols (2) as a function of temperature.

than liberation of heat, as is the case for ordinary chemical bonds; breakage of such a bond is accompanied by evolution of heat. Taking into account the endothermic character of the breakage of hydrophobic bonds, we obtain the graph shown in Fig. 4 for the temperature function of the enthalpy of a system in which bond rupture occurs at a temperature T_1. We must again emphasize that the ordering effect of the nonpolar groups on the water is intensified when the hydrophobic bonds are broken. At the same time, the number of hydrogen bonds in water is known to decrease substantially more rapidly with rising temperature in systems where these bonds are stabilized by nonpolar molecules than in water. This can be seen from Fig. 5, which shows the number of H bonds in pure water and in the presence of aliphatic hydrocarbons as a function of temperature [4].

Hence it follows that the heat capacity of a system containing water ordered by the influence of nonpolar groups must increase with the ordering effect of these groups, since the more extensive water structures "fuse" in this case.

We can thus conclude that the slope of the enthalpy curve in Fig. 4 should be sharper above the transition temperature T_1 and that the negative change in enthalpy should consequently be smaller when this transition occurs at a higher temperature.

That the endothermic character of hydrophobic-bond rupture decreases with rising temperature follows directly from the mechanism by which the hydrophobic interaction arises: the ordering influence of the nonpolar groups on the water decreases as the temperature rises and the enthalpy term, which is determined by the hydrogen bonds formed in the water during ordering, is therefore reduced.

According to our microcalorimetric data and the indirect data of various authors (see also [5]), however, the experimentally measured enthalpy of denaturation for different macromolecules is positive. There must consequently be still another mechanism for stabilization of the native structure of macromolecules, making a positive enthalpic contribution that exceeds the negative contribution of the hydrophobic bonds, and determining the enthalpic character of the denaturation process. This structure-stabi'izing factor evidently comprises all the bonds broken during denaturation. Of prime importance among these are the hydrogen bonds present in large numbers in the macromolecule. Designating the positive enthalpic contribution as ΔH, we obtain the following equation for the total change in enthalpy during denaturation:

$$\Delta H = \Delta H_{\text{н}} - \Delta H_{\varphi}.$$

As can be seen from Fig. 4b, ΔH should be some (linear in first approximation) increasing function of temperature. Let us examine the term ΔH_H in greater detail and determine the extent to which it can be treated as a constant.

It is generally assumed that when an H bond in a macromolecule ruptures an exchange reaction takes place and the bond between the groups is replaced by two bonds to water molecules. Actually, ΔH is thus the difference in the enthalpies of formation of H bonds between the polar groups of the macromolecules and between the water molecules on the one hand and between the polar groups and the water molecules on the other hand.

Following Shellman, it was long believed that ΔH reached a value of the order of 1500 cal per mole, so that hydrogen bonds were considered to be the principal factor stabilizing native macromolecular conformation. However, it was subsequently demonstrated that the exchange reaction that takes place during H-bond rupture leads to almost complete compensation of the factors constituting ΔH_H and that the gain in enthalpy is close to zero, so that it cannot play any decisive role in maintaining macromolecular structure. This role has therefore come to be more and more fully ascribed to hydrophobic interactions. According to Tanford, hydrophobic bonds alone can provide complete macromolecular stability. However, this extreme viewpoint can scarcely be valid, if for no other reason than the observed sign of the change in enthalpy during denaturation mentioned above.*

Moreover, it seems highly improbable that the exchange reaction involved in H-bond rupture goes to completion over the entire temperature range. Actually, the hydrogen bonds in the macromolecule form a cooperative system, which breaks down only as a unified whole. At the same time, H-bond formation with water should proceed in a substantially less cooperative manner and depend to a large extent on the structural temperature of the water. Hence it is clear that, while an almost complete exchange reaction occurs when the H bonds in the macromolecule are broken at low temperatures, where the probability of H-bond formation in the water and thus with the water is high, the completeness of the exchange must decrease as the structural temperature rises, i.e., the value of ΔH_H, which is a difference effect, must increase.

This rise in ΔH with temperature is especially pronounced in systems where the H bonds in the water are artificially stabilized by the presence of other molecules. As has already been noted, the decrease in the number of H bonds in the water with rising temperature is substantially more rapid in this case. Judging from the calculations made by Nemethy and Scheraga, a rise in temperature from 20 to 70° should cause the number of H bonds in aliphatic hydrocarbon solutions to decrease by about 15%. It is therefore to be expected that ΔH_H will increase by the same amount over this temperature range.

We can thus conclude that both constituents of the enthalpy of denaturation should cause a rise in this factor with temperature. The functional dependence of ΔH_H on temperature results from the most general properties of water and should therefore be universal in character.

LITERATURE CITED

1. P. L. Privalov, Biofizika, 8:308 (1958).
2. P. L. Privalov, K. A. Kafiani, and D. R. Monaselidze, Biofizika, 10:393 (1958).
3. J. F. Brandt, J. Am. Chem. Soc., 86:4291 (1964).
4. G. Nemethy and H. Scheraga, J. Chem. Phys., 36:3401 (1962).
5. B. N. Sukhorukov, Yu. Sh. Moshkovskii, T. M. Birshtein, and V. N. Lystsov, Biofizika, 8:294 (1963).

---

*A whole series of structural data also indicate that a material contribution is made by both hydrophobic and hydrophilic interactions to the structural stability of proteins and other biopolymers [for example, see: The Structure of Hemoglobin and Lysozyme; J. C. Kendrew et al., Nature, 160:669 (1961); Fillips et al., Nature, 220:23 (1966)].

STRUCTURAL ROLE OF WATER IN FIBRILLAR PROTEINS AND POLYPEPTIDES

N. G. Esipova and Yu. N. Chirgadze

Institute of Biological Physics
Academy of Sciences of the USSR

The problem of the interaction of water with proteins, which is a component of that of the structural equilibrium of the solvent—macromolecule system, is usually divided into two separate questions: the influence of water on macromolecular structure and the state of the water itself around this structure.

In essence, the very fact that a molecule dissolves in water indicates that a definite number of polar groups are in contact with the solvent, while the hydrophobic ring usually has no contact with the ambient medium [1]. However, if the number of polar groups at the surface is sufficiently large to hold the molecule in solution, some of the hydrophobic groups may be in contact with the solvent. A number of authors have hypothesized that ordering of the solvent is possible around hydrophobic sections of the surface (e.g., see [2]).

The possibility of the structuring of water in biological systems has often been discussed in the literature. We can make a more or less definite judgment regarding this problem from the results yielded by x-ray diffraction analysis of protein systems (both globular and fibrillar).

The myoglobins of a number of animals, hemoglobin, and lysozyme are among the globular protein structures that have been investigated in detail. It has been found that most of the hydrophobic groups in myoglobin and hemoglobin are in the interior of the molecule and cannot come into contact with water [3, 4]. The polar residues are distributed in such fashion that they cannot promote ordering of water. However, the problem of the structuring of large amounts of water (in the so-called "water coat") is still a matter of dispute, because of the lack of exhaustive experimental evidence. For example, some of the nonpolar residues in lysozyme are located at the surface of the molecule, so that we cannot exclude the possibility that water is ordered around these groups [5].

The data for fibrillar systems are far more complete. In addition, these systems correspond more closely to intact cellular structures: only mild treatment of cellular material makes it possible to investigate their structure without any danger of impurity effects. Specimens are prepared in the form of films or fibers and the genesis of their formation from a given solvent can shed some light on the role of water in such systems.

All fibrillar proteins can be divided into two groups: those whose configuration is governed by the backbone of the polypeptide chain and those whose specific configuration is produced by regular alternation of definite amino acid residues [6].

The α-helix is considered to be the most stable configuration for the polypeptide backbone. In this configuration, a system of CO—NH hydrogen bonds runs down the outside of the helix and parallel to its axis [7], so that the α-helix is unstable in polar solvents, including water. One feature of the α-helical configuration is the fact that most of the amino acids, including those that are polar and hydrophobic, will fit into it.

Investigation of the melting of poly-γ-benzyl-l-glutamate in benzene is interesting in this connection. It was shown by Doty and Yang [8] that, in contrast to other polypeptides, the phase transition occurs when the temperature is reduced rather than when it is raised (which would be understandable in terms of a decrease in the stability of the hydrogen-bond system). One possible explanation of this effect is that the ordering of the solvent molecules around the benzene rings makes the entire system entropically disadvantageous at low temperatures, thus causing breakdown of the ordering in the macromolecule itself.

The stability of fibrillar systems in aqueous solutions is evidently governed principally by the possibility of aggregation of the hydrophobic amino acid residues of different molecules, so that contact with the solvent is made by the polar side chains (as is the case for a true solvent), since strong hydration of these chains should reduce the possibility of breakage of the CO—NH hydrogen bonds (which have no energy advantage over bonds between the corresponding groups and the water). It can further be asserted that the configuration of these complexes must, in order to provide sufficient stability, prevent contact between the water and the CO$=$NH hydrogen bonds. The so-called "spiral helix" that occurs in myosins and tropomyosins insulates these bonds from contact with water [9].

The second class of proteins comprises the various silk fibroins and collagens.

The specific configuration of the polypeptide chain of a silk fibroin is determined either by large numbers of glycine and alanine residues (*Bombyx mori*) or by a single alanine residue (*Tussah*). It has been proved by x-ray diffraction analysis that the corresponding polypeptides and proteins are isomorphic [10]. These polypeptides consist of hydrophobic amino acid residues and water-structuring effects around the hydrophobic groups are therefore to be expected in the system.

However, silk fibroin is insoluble in water and, on being transferred to LiBr solution, loses its characteristic extended chain configuration, folding into an α-helix [11]. Poly-l-lysine, which forms a classical β-structure, also passes into the α-helical configuration at high moisture contents (84%) [12]. It is therefore understandable that any discussion of the role of water loses most of its meaning for proteins in the β-configuration.

Proteins of the collagen class are a special case with respect to the importance of hydration.

The structure of these proteins results from the presence of large amounts of imino acids and glycine. Imino acids lack NH groups, so that linking of all the peptide groups into systematic hydrogen-bond networks is impossible. Synthesis of model polymers, which was first carried out by Shibnev and Debabov [13-15], showed that it is possible for the specific callogen configuration to arise in systems where, firstly, only one systematic hydrogen-bond network can be formed and, secondly, where the presence of hydrophobic side chains creates suitable conditions for structuring of water, as in (glycyl—prolyl—proline)$_n$. We used these systems as a basis for a study of the characteristics of the interaction of water with proteins of the callogen group. Such studies can shed some light on the amount of water that can be held in a stable state around a configuration with hydrophobic residues in the lateral voids and thus on the size of the ice-like coat, if one exists.

During 1955–1958, we conducted sorption, x-ray diffraction, and spectroscopic investigations of collagen and its soluble component, procollagen [16, 17].

We investigated the intensity of the interferences characterizing the ordered regions in collagen as a function of water content. It was found that, when the moisture was completely removed (the specimens were completely dried over P_2O_5 and $CaCl_2$ or in a vacuum), the intensity of the main interference with $d \approx 2.9$ Å, which is characteristic of proteins of this group, fell to zero. Calculation of the Fourier transformants for a "dry" collagen molecule yielded a high value for the (0010) reflex. This enables us to conclude that water is a necessary component of the collagen structure, its absence leading either to disordering or to complete destruction of the specific configuration.

A more detailed analysis of the data showed that the interference with $d = 2.9$ Å increased more rapidly than the equatorial interference with $d = 10$ Å. We attributed this phenomenon to the fact that water regularly incorporated into the structure scatters in phase with the groups giving a projection of about 2.9 Å on the axis of the helix. It is most probable that the water is regularly attached to the $C = O$ groups of the imino acids, which do not enter into the systematic hydrogen-bond network, thus forming additional hydrogen bonds that stabilize the chain configuration.

These data were quite recently confirmed in studies of collagen hydration conducted by other methods.

Investigation of the arrangement of the water in collagen by the nuclear-magnetic-resonance method showed that it forms regular structures with a preferential period of 28.6 Å, i.e., one corresponding to the main period in collagen [18].

Investigation of the infrared absorption spectra of procollagen films in turn established that the intensity of the peptide bands increases on hydration; this indirectly confirms our conclusions [19].

In analyzing the formation mechanisms of the left-handed helix in polyproline and certain proteins of the collagen class, Harrington et al. concluded that water is, in general, a necessary medium for development of the levorotary configuration [20]. The problem lies in whether water is merely a plasticizing agent that promotes development of this configuration in polypeptide macromolecules, whether it can be a supplemental stabilizing agent, or whether it is a necessary structural element in the collagen configuration. Our data on the disappearance of the interference with $d = 2.9$ Å on sufficiently complete dehydration of collagen supports the second or third hypothesis, since, if water merely promoted formation of collagen and did not participate in its structure, the latter would not be deformed on drying.

The spectroscopic data obtained by Blout [21] for specimens of polyproline II in the levorotary configuration did not reveal bound water molecules.

In order to resolve these problems for structures of the collagen class, we undertook an investigation of the conditions for stability of the specific configuration in nonaqueous solvents. Procollagen prepared by the technique suggested by Orekhovich and Shpikiter [22] was dissolved in ethylene glycol, ethylene chlorohydrin, and propylene glycol.* The resultant homogeneous solution was subjected to complete "fusion" (thermal denaturation) and preparations similar to the cold gelatin produced from aqueous procollagen solutions were made up from the specimen. In view of the known decrease in the transition temperature of procollagen in nonaqueous solvents, "cold" gelatin was also prepared in the same manner at low temperatures. The dry

---

*It should be noted that the nonaqueous solvents employed were polar.

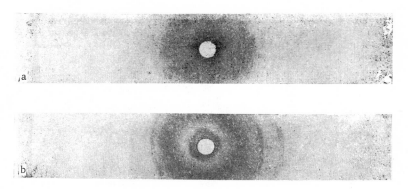

Fig. 1. X-ray patterns of "cold" gelatin produced by thermal denaturation of collagen. a) Denaturation in nonaqueous solvent (specific collagen diffraction not observed); b) denaturation in water.

Fig. 2. Restoration of characteristic diffraction pattern of collagen after moistening of gelatin denatured in nonaqueous solvent.

Fig. 3. X-ray pattern of water-containing (glycyl−prolyl−oxyproline)$_n$ specimen.

Fig. 4. X-ray pattern of anhydrous (glycyl−prolyl−oxyproline)$_n$ specimen.

"cold" gelatin specimens were subjected to x-ray diffraction analysis. Figure 1 shows a typical x-ray pattern of such a specimen. As can be seen, it exhibits no signs of specific collagen diffraction. An x-ray pattern for cold gelatin produced by thermal denaturation of collagen in water is also shown for comparison.

The disappearance of the specific configuration after nonaqueous thermal denaturation indicates that water is not merely a necessary supplemental stabilizing agent but that it is a specific structural element of the collagen helix. It is interesting that addition of a drop of water to the "hot" gelatin can completely restore the specific collagen configuration. Figure 2 shows the x-ray pattern of a collagen specimen restored by moistening, indicating complete renaturation of the specific collagen configuration.

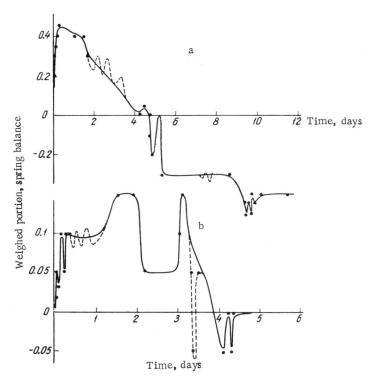

Fig. 5. Sorption kinetics of two (glycyl−prolyl−oxyproline)$_n$ specimens. a) Containing water; b) not containing water.

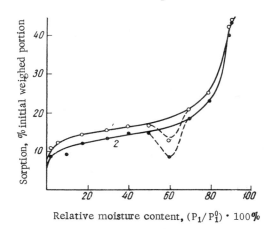

Fig. 6. Sorption curves for two (glycyl−prolyl−oxyproline)$_n$ specimens. 1) Anhydrous; 2) specimen containing water.

We thus concluded that water plays a special structural role in collagen and is necessary for formation of its specific configuration.

Let us now consider the important problem of the number of water molecules that must be present in a collagen-type configuration and the amount of water retained by the structure. It should be noted that additional structuring is possible around the hydrophobic imino acids. It follows from the sorption curves for procollagen that the molecule retains about 20 g of water per 100 g of protein [16]. Collagen is known to be very rich in polar amino acids, whose sorptive capacity is rather high. It might be suggested that precisely these groups must be blocked by water for the normal type of collagen structure to develop. This hypothesis is refuted, however, by the fact that the intensity of the interference with d = 2.9 Å increases to a greater extent than the 10-Å reflection during hydration and by nuclear-magnetic-resonance data indicating that the water has an ordered arrangement conforming to the specific structure, i.e., it occurs in the segments rich in imino acids and glycine. Nevertheless, the quantitative distribution of the water between the two regions is to a large extent governed by the sorptive capacity of the polar groups themselves and by the amount of structured water that can be held by the nonpolar groups.

Fig. 7. X-ray pattern of (glycyl−prolyl−oxyproline)$_n$ specimen crystallized during sorption (at relative moisture content of 60%).

Fig. 8. X-ray pattern of crystalline (glycyl−prolyl−oxyproline)$_n$ specimen.

We employed polymers of the (glycyl−amino acid−imino acid)$_n$ type as models of the ordered region of collagen and investigated the influence of water on their structure, making sorption and x-ray diffraction analyses. Two specimens of (glycyl−proline−oxyproline)$_n$ were studied in the sorption experiments. Figures 3 and 4 show their x-ray patterns. As can be seen, one of them was in a specifically collagenic configuration and had the set of interferences characteristic of it, while the other was completely amorphous. No water was used in the synthesis of this second specimen.

We investigated both the sorption−desorption isotherm and the curves representing sorption kinetics for the aforementioned specimens.* The total amount of water absorbed during the first stage of sorption differed for the dehydrated and undehydrated specimens. Their sorption isotherms and sorption kinetics also differed. The latter were not the same as those usually observed in polymer specimens. Instead of an asymptotic curve reaching saturation after a certain time, the (glycyl−prolyl−oxyproline)$_n$ specimens sorbed and then released water when a definite portion of vapor was added to a MacBain apparatus. Especially large fluctuations were observed near a relative humidity of about 60% (Figs. 5 and 6). The appearance of the dehydrated specimen changed at this relative humidity, turning from a white flaky powder to a yellow substance resembling a mixture of finely crystalline structures. Figure 7 shows the x-ray pattern of this specimen, which is identical to those of crystalline (glycyl−prolyl−oxyproline)$_n$ specimens (Fig. 8). The second specimen subjected to sorption experiments also crystallized during moistening, but the fluctuations were substantially smaller in this case. The crystallization reduced to achievement of better packing.

The most characteristic sorption detail was the fact that the water content of the specimen was lower after crystallization than during the preceding sorption phase. We thus got the impression that the actual crystallization process requires additional water, while a crystallized specimen can retain only a considerably smaller amount. Considering that the side chains of the (glycyl−prolyl−oxyproline)$_n$ polymer contain principally hydrophobic groups (except for the peptide C = O and oxyproline OH), it can be presumed that the additional water promoting crystallization of the polymer is itself ordered around the hydrophobic groups, but this process is entropically disadvantageous and the system releases the excess solvent. This would explain

* The sorption experiments were conducted in conjunction with T. V. Gotovskii.

the unusual trend of the sorption—desorption isotherms for the polymers. However, it is quite possible that we are dealing with a specimen always in a nonequilibrium state, so that the aforementioned curve cannot be interpreted as an isotherm in the strict thermodynamic sense.

Polymers isomorphic to collagen thus require water for crystallization, but the crystalline polymer in the equilibrium state can retain only a small number of water molecules (no more than three) per residue, i.e., considerably fewer than the true collagenic (noncrystalline) polymer configuration. The polymer retains the same number of water molecules when the moisture content is further raised, sorption increasing only when it completely dissolves. One gets the impression that the crystal shape is characterized by a strictly constant number of water molecules. Finally, it is obvious that this amount of water (which is structural) cannot form any sort of ice-like coat.

All other polypeptides isomorphic to collagen also require small amounts of water for formation of specific configurations [crystalline (glycyl—prolyl—proline)$_n$, (glycyl—prolyl—alanine)$_n$, and (glycyl—oxyprolyl—oxyproline)$_n$] and undergo additional crystallization during sorption.

The sorption of water by these polypeptides can be divided into two stages: 1) development of a typically collagenic structure; 2) complete crystallization of the polymer, with a small amount of water in each cell.

If we compare the data obtained for the model polypeptides and for collagen and procollagen, it becomes obvious that most of the water sorbed by the natural protein goes to the polar side chains. The presence of polar regions that modify the overall thermodynamic balance of the system leads to formation of an equilibrium collagen structure capable of existing in tissues. The amount of water that goes to the regular hydrophobic areas is small, so that formation of a structured coat is impossible. Nevertheless, this small number of water molecules bound to the hydrophobic groups plays a decisive role in formation of the specific collagenic conformation. In the case of polypeptides, water permits further ordering of the macromolecules, which leads to formation of a highly crystalline structure.

In conclusion, we will consider different aspects of the problem of water and fibrillar proteins, drawing conclusions from the material presented above.

After the elementary initial premise that "the contents of the cell are in aqueous solution, which means that this water is important," we are faced with a series of specific problems regarding the state of the water and its role in structure formation.

We have shown that it is scarcely possible to assume development of a large structured-water shell around the hydrophobic groups during formation of the secondary structure of fibrillar proteins.

The need for water in building up macromolecular structure has been explained in different ways for fibrillar proteins of different types.

The role of water in the structure of α-helical proteins is negative, if one can make such a statement. The α-helix is unstable in an aqueous medium (disregarding special homopolypeptides), so that the molecules are "forced" to aggregate and form a spiral helix. On the other hand, the collagen configuration has maximum stability in an aqueous medium; any material dehydration leads to severe disordering, i.e., to breakdown of the characteristic structure. The water in collagen, which is a necessary medium for formation of the so-called collagenic configuration, is also an obligatory structural element.

In considering the formation of biological structures at the submolecular level, we must discuss the combination of macromolecules into larger fibrillar aggregates in an aqueous medium.

Retention of the linearity of each constituent molecule is a necessary condition for construction of a fibrillar tissue; on the other hand, correct alignment of individual segments of the molecules with one another is important. The first condition can be satisfied by massing of the polar amino acid residues in individual sections of the chain, so that the electrostatic repulsion of the charged groups keeps the polyelectrolyte chain in the extended state.

Proper aggregation of a tissue requires some quite definite regularity in the arrangement of the chemical groups along the chain. The simplest way to obtain such regularity is proper alternation of polar and nonpolar segments. The requirement that there be clusters of polar groups in definite areas is apparently a general one in the aggregation of fibrillar protein molecules. This also ensures the best contact among the hydrophobic groups. The requirements for regularity of chemical structure may be more stringent if the proteins aggregate (e.g., the proteins of the actomyosin complex).

Each new "stage" of structure formation thus imposes specific requirements on the amino acid sequence. While every third residue must be of a definite type for a stable secondary structure with a three-chain spiral helix and every seventh residue for a twin cable of α-helices, a stable fibrillar tissue requires spatial separation of the polar and nonpolar groups. Regions with hydrogen-bonded water and reduced density may be formed in this case.

In conclusion, it should be emphasized that the purpose of this paper was to present actual experimental data on the structural role of water in fibrillar proteins and conclusions following from them. At present, this approach certainly seems to be the only fruitful one, since sufficient research has not as yet been done on the properties of water itself. Conclusions based on *a priori* considerations (like the tendency toward formation of hydrogen-bond networks) are therefore often ill-grounded and do not agree with experimental results.

I am deeply grateful to N. S. Andreeva for her valuable advice and constant support in this work.

LITERATURE CITED

1. T. M. Birshtein, present volume, p. 9.
2. I. M. Klotz, Federat. Proc., Vol. 24, Suppl. No. 15, 24 (1965).
3. J. C. Kendrew et al., Nature, 190:669 (1961).
4. A. F. Gullis, H. Muirhead, A. F. Perutz, and M. G. Rossman, Proc. Roy. Soc., A265(15): 161 (1962).
5. C. C. Blake, A. C. North, D. C. Phillips, et al., Nature, 206:757 (1965).
6. N. S. Andreeva, Vysokomolek. Soed., 1:308 (1959).
7. L. Pauling and R. Corey, Proc. Nat. Acad. Sci., 37:235 (1951).
8. P. Doty and J. Yang, J. Am. Chem. Soc., 78:498 (1956).
9. M. I. Millionova, Biofizika, 2:157 (1957).
10. Go. Yukichi, Noguchi Junzo, Masamoto Asai, and Tadao Hajakawa, J. Polymer Sci., 21: 147 (1956).
11. N. S. Andreeva and M. I. Millionova, Biofizika, 3:28 (1956).
12. U. Schmueli and W. Traub, J. Mol. Biol., 12:205 (1965).
13. V. A. Shibnev and V. G. Debabov, Izv. Akad. Nauk SSSR, Seriya Khim., p. 1043 (1964).
14. V. A. Shibnev, V. N. Rogulenkova, and N. S. Andreeva, Biofizika, 10:164 (1965).
15. V. A. Shibnev, K. T. Poroshin, and V. S. Grechishko, Izv. Akad. Nauk SSSR, Seriya Khim., p. 1493 (1966).
16. N. G. Esipova, N. S. Andreeva, and T. V. Gotovskaya, Biofizika, 3:529 (1958).
17. V. N. Rogulenkova, V. A. Shibnev, and N. S. Andreeva, Biofizika, 4:35 (1963).
18. H. J. C. Berendsen, J. Chem. Phys., 36:3297 (1962).
19. Yu. N. Chirgadze, S. Yu. Ven'yaminov, and S. L. Zimont, present volume, p. 51.

20. W. F. Hurrington and P. H. von Hippel, Adv. in Prot. Chem., 16:1 (1961).
21. E. J. Blout and J. D. Carver, in: Collagen (edited by Ramachandran), London (1966).
22. V. N. Orekhovich and V. O. Shpikiter, Biokhimiya, 20:438 (1955).

INVESTIGATION OF THE STATE OF WATER IN THE STRUCTURE OF PROTEINS AND POLYPEPTIDES BY INFRARED SPECTROSCOPY

Yu. N. Chirgadze, S. Yu. Ven'yaminov, and S. L. Zimont

Institute of Biological Physics, Academy of Sciences of the USSR
and Physicotechnical Institute

The infrared spectrum of a substance simultaneously reflects the molecular vibrations of all the components comprising it. When a protein contains water, its absorption bands are therefore superimposed on the absorption spectra of the polypeptide chain and of the side and terminal groups of the biopolymer molecule. The minimum amount of water that can be detected in a dried protein from its infrared spectrum is about 0.5% by weight. The minimum amount of protein whose infrared spectrum can be studied in aqueous solution is several percent by weight. Although structural water is manifested in one way or another in the spectrum of almost any polypeptide or protein, very little research has been done on the hydration of biopolymers. The present work was undertaken to study the possibility of determining the content and characteristics of the structural water in polypeptides and fibrillar proteins from their infrared spectra. All the experimental illustrations are for procollagen, which is being studied in our laboratory.

Fundamental-Normal-Vibration and Overtone

Regions of the Water Spectrum

Hydration and a number of other phenomena in polypeptides have been studied with both ordinary water H_2O and heavy water D_2O (Fig. 1). In working with thin layers of heavy water, it is difficult to avoid exchange with the normal water present in the atmosphere; the spectra presented therefore also exhibit additional peaks caused by vibrations of molecules of the HDO type (these are indicated by dots). The triatomic water molecule has three types of normal oscillations in the fundamental-vibration region: two types of valence vibrations (asymmetric and symmetric) and one type of deformation vibration. We are obliged to deal with water molecules that always participate in forming hydrogen bonds of the OH...O (OD...O) type. As a result, the frequencies are somewhat different from those for the gaseous state, the intensity of the valence vibrations is greatly increased, the band is broadened, and we cannot distinguish the two types of valence vibrations in the complex curve. The absorption of water is only one-tenth as great in the overtone region, but two peaks near 5150 and 6900 cm^{-1}, produced by composite vibrations of the valence and deformation types, are clearly visible [1].

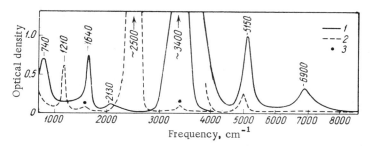

Fig. 1. Infrared spectrum of liquid water in fundamental-vibration region (700-4000 cm^{-1}) for layer about 4 μ thick and in overtone region (4000-8500 cm^{-1}) for layer about 50 μ thick. 1) Spectrum of ordinary water H$_2$O; 2) spectrum of heavy water D$_2$O; 3) bands belonging to impurity water of HDO type.

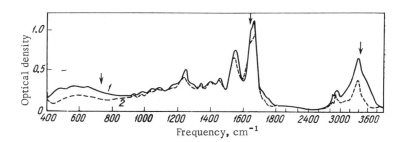

Fig. 2. Fundamental-vibration region of infrared spectrum of procollagen film several microns thick at different humidities. 1) 100% humidity; 2) 0 humidity (over P$_2$O$_5$). The arrows indicate the regions in which the structural water absorbs.

Manifestation of Water in the Procollagen Spectrum

Figure 2 shows the fundamental-vibration region of the spectra of procollagen films in two different states: at relative humidities of 100 and 0% ordinary water. It can be seen that there are material changes in the spectral curve, consisting principally in an increase in absorption for almost all the bands observed. However, it is impossible to draw definite conclusions regarding the change in peptide absorption for the series of fundamental amide bands (with maxima near 1650 and 3330 cm^{-1}), since the valence and deformation vibrations of water absorb in almost the same regions. Use of heavy water makes interpretation more difficult, since, in addition to replacement of H$_2$O molecules by D$_2$O, some of the peptide groups in the amorphous regions of the polymer and in the side chains of the amino and imino acid (oxyproline) residues undergo substitution (NH and OH being replaced by ND and OD). Analysis of the intensity of the band near 1550 cm^{-1} (amide II) permits us to conclude that there is an increase in the intensity of peptide absorption. It has been found that when the humidity is raised from 0 to 100%, the band intensity rises somewhat, until a humidity of 60% is reached, and then remains constant. This indicates that the structural characteristics are not altered when the humidity is further raised. This aspect of carbonyl absorption can be studied only after the water peak near 1640 cm^{-1} has been eliminated, which can be done by using heavy water. It is obvious from the foregoing that the fundamental-vibration region provides us with information on the properties of the peptide group, i.e., on the biopolymer molecule itself, while the water is manifested principally as an "interference factor."

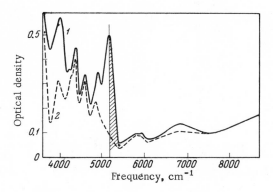

Fig. 3. Overtone region of absorption spectrum of procollagen film 250 μ thick at different humidities. Symbols as in Fig. 2.

The situation is totally different in the overtone region (Fig. 3). The absorption bands for the polypeptide molecule, which are produced by the composite CH and amide-group vibrations, cover the range from 3900 to 5000 cm^{-1}, while the water peak exhibits an absorption maximum at about 5150 cm^{-1}; no other bands are present in the latter region.

The absorption in the higher-frequency region is due to the second overtones, which are far weaker and broader, and hence of less interest. It thus appears possible to make separate studies of the structural water in proteins. For example, we can determine the hydration curve by measuring the intensity of the 5150 cm^{-1} peak at different humidities (the maximum intensity can be used, but the integral intensity is better characterized by the magnitude of the shaded portion of the band, as shown in Fig. 3). We obtained a result for procollagen similar to those yielded by other sorption-curve measurement methods. Use of heavy water leads to complete disappearance of the 5150 cm^{-1} peak (which is displaced into the low-frequency region, thus indicating that almost all the water observed in the spectrum is replaced; the actual figure is about 97%, taking into account the accuracy of the method in this region).

Subtraction of the spectrum at zero humidity from the spectrum at 100% humidity yields a curve identical to the spectrum of the water layer, i.e., no change in the intensity of the amide bands of the composite vibrations occurs during hydration. This fact does not, however, contradict the data obtained for the fundamental-vibration region, because of the more complex dependence of the composite bands on the electrooptical parameters of the peptide group.

The overtone region gives us an opportunity to study the orientation of the water molecules from the dichroism of the 5150 cm^{-1} peak. The perpendicular dichroism of this band in the collagen spectrum [2] indicates that most of the water molecules are arrayed perpendicular to the axes of the fibrils, forming intermolecular hydrogen bonds. We found that the dichroism of this peak in the tail tendon of the rat increases with humidity and reaches its maximum at a humidity of about 60%; the maximum amount of protein is in the ordered "crystalline" form at this same humidity. For a number of reasons, determination of the crystalline portion of a polymer is best carried out with the absorption bands in the overtone region [3].

Conclusions

We have thus shown that the most convenient spectral region for study of structural water by infrared spectroscopy is the overtone region, where the water has a separate absorption peak. The principal difficulty lies in the fact that the different types of structural water (intermolecular, intramolecular, superficial, etc.) apparently differ little in their spectral characteristics, so that they are all manifested in the same absorption peak near 5150 cm^{-1}. On the other hand, the change in the properties of the polypeptide molecule itself during hydration is best observed in the amide bands in the fundamental-vibration region, although the features noted above must be taken into account.

LITERATURE CITED

1.   M. A. El'yashevich, Atomic and Molecular Spectroscopy [in Russian], Moscow (1963).
2.   R. D. B. Fraser and T. P. Macraae, Nature, 183:179 (1959).
3.   R. D. B. Fraser and T. P. Macraae, J. Chem. Phys., 28:1120 (1958); 29:1024 (1958); 31:122 (1959).

INVESTIGATION OF THE STATE OF TISSUE WATER BY INFRARED SPECTROSCOPY

A. I. Sidorova, I. N. Kochnev, L. V. Moiseeva, and A. I. Khaloimov

A. A. Zhdanov Leningrad State University

As a result of the rapid development of infrared spectroscopy in recent years, the optical spectrum of water in its three principal aggregated states has been thoroughly studied. The vibration spectrum of an isolated water molecule has been calculated theoretically and measured experimentally [1]. The spectra of liquid water and ice have also been investigated in great detail [2]. The reliability of the frequencies obtained has been repeatedly verified by allied methods: the data yielded by infrared spectra have been compared with those obtained for combined-scattering spectra, cold-neutron scattering, nuclear magnetic resonance, dielectric losses, etc. However, the interpretation of the vibration spectrum of water and the assignment of certain frequencies are not yet completely unambiguous [3], although use of isotopic substitution has substantially facilitated deciphering of the spectrum.

The numerous reliable experimental spectroscopic data have aided in introducing some order into the diverse theories of the structure and properties of water and aqueous solutions [4]. Finally, the broad application of computers characteristic of the past 5–10 years has enabled a number of authors to reduce their theoretical hypotheses about water structure to numerical results and to verify them experimentally.

In our opinion, all these factors lend timeliness to our attempt to deal with the traditional hazards in consideration of the decisive role of water in the infrared spectra of solutions, particularly those of biological objects. On the other hand, the level of current knowledge about water permits study of a biological object in a condition similar to its native state, i.e., in the presence of water, which is the sole medium for and an active participant in vital processes.

We have previously called attention to the fact that the infrared water-absorption bands in the spectrum of almost any biological tissue from a vertebrate (the brain, skeletal muscles, kidneys, myocardium, blood, blood serum, etc.) are, in first approximation, identical to the spectrum of ordinary pure water [5]. The typical contour of a number of pure-water bands is reproduced quite well: thus, for example, the valence absorption band at 3400 cm^{-1} is always accompanied by the additional maximum at 3280 cm^{-1} characteristic of the symmetric OH vibration of the water molecule.* Similarly, the 5200 cm^{-1} band displays the additional maximum

---

*The 3400 cm^{-1} band is reproduced particularly accurately in the brain, kidneys, and blood serum.

Table 1. Vibration Spectrum of Liquid Water

$\nu_1+\nu_2+\nu_3$	$\nu_1+\nu_3$	$\nu_2+\nu_3$	$\nu_1,\nu_3,2\nu_2$	$\nu_2+\nu_L$	ν_2	ν_L	ν_{h_2}	ν_{h_1}	Meas. unit
8500	6900	5200	3400	2130	1645	740	175	60	cm^{-1}
5000	1000	300	2	70	5	3	50	—	$m\mu$

at 5160 cm^{-1} characteristic of pure water. It is therefore possible to conduct special investigations of the tissue-water absorption bands of biological tissues and to extract information on the structure and behavior of this water from detailed analysis of the bands.

In the present article, we will consider several techniques that have been developed in our laboratory for spectral study of the state of water and water metabolism in the living organism.

This work was undertaken at the urging of N. A. Verzhbinskaya and she has participated actively in it, in connection with her research on the hematoencephalic barrier [6].

Table 1 presents the vibration spectrum of liquid water. The top line shows the generally accepted frequency assignments [7], the second line shows the approximate positions of the absorption-band maxima at room temperature, and the bottom line shows the thickness of the layer required to produce the proper absorption-band contour.

The figures in the bottom line are of great importance for investigating intact biological tissues in a condition as close as possible to their native state.

The principal, so-called fundamental frequencies in the water spectrum are the symmetric and asymmetric valence vibrations ν_1 and ν_3, the deformation vibration ν_2, and the liberation vibration ν_L, very thin layers (2-3 μ) being required to produce the proper band contours. It is very difficult to prepare absolutely plane-parallel layers of undamaged biological tissues having this thickness. Nevertheless, as will be seen from the material that follows, some of our investigations were conducted on the fundamental absorption bands.

The frequencies of the translational vibrations of water molecules ν_{h_1} and ν_{h_2}, which are associated with hydrogen bonds, lie in the far-infrared region and we are not concerned with them at present.

The most favorable conditions for obtaining sharp absorption-band contours for water in biological tissues and for quantitative interpretation of the bands occur in the near-infrared region, between 4000 and 7000 cm^{-1}, which contains the absorption bands corresponding to the overall molecular-vibration frequencies and where the thickness of the tissue specimen can reach 3 mm (for the 5200 and 6900 cm^{-1} bands).

The $\nu_2 + \nu_L$ band at 2130 cm^{-1} occupies an intermediate position and we have repeatedly used it, since it is convenient for experimentation and requires a layer thickness of 70 μ.

Let us now move directly on to a description of spectroscopic methods for studying water.

Quantitative Evaluation of Tissue-Water
Exchange Rate

This method was developed in our laboratory, in conjunction with N. A. Verzhbinskaya and has been used in many experiments, being principally of biological significance.

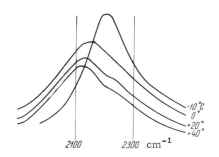

Fig. 1. Temperature displacement of deformation-liberation band maximum for water in cerebral tissue.

It is based on the fact that the rate of the isotopic exchange D $\rightleftharpoons$ H is considerably higher in OH bonds than in NH or CH bonds.

It is assumed that heavy water injected into an animal's blood stream participates in deuteron exchange only with tissue-water molecules and that, by establishing the number of OD bonds in a tissue, we can determine the rate of water exchange. This hypothesis is apparently valid, since the data obtained by our method agree with those previously obtained by Verzhbinskaya with the phosphorus isotope P^{32} [6].

We attempted to cover the initial processes of deuteron exchange in the body as thoroughly as possible and to determine the rate of water exchange before establishment of isotopic equilibrium. However, the shortest feasible deuteron-exchange time was one minute.

An animal was injected with heavy water (D_2O) in a dose corresponding to 0.5% of its total body weight and kept alive for a definite time. The amount of heavy water that entered the tissues was evaluated by measuring the ratio of the maximum heights of the absorption bands for heavy water at 2500 cm^{-1} (valence vibration of the OD bond of the HDO molecule) and for light water at 2130 cm^{-1} (the deformation-liberation band of the H_2O molecule). On the basis of several hundred experiments, we estimated the error in determining the D_2O content of the tissues to be 10-15%.

The results of these measurements are conveniently presented in the form of graphs, with the time after injection of the heavy water (1, 2, 3, 5, 10, . . . , 100 min, etc.) plotted along the abscissa and the percentage D_2O content in the ordinary water of the biological tissue in question plotted along the ordinate.

These curves represent the rate at which the D_2O enters the tissues of the brain, muscles, kidneys, etc., from the blood; it is governed by the specific characteristics, structure, and metabolic rate of the tissue, i.e., by biological parameters. A substantial part of the experimental error can be eliminated by recalculating the D_2O content of the tissue determined from the spectrum in terms of the D_2O content of the animal's blood serum. This conversion yields a high degree of standardization of the initial experimental conditions.

The details of this technique have been described in previous works [8].

We conducted extensive investigations of the changes in water-exchange rate in many vertebrates. Different tissues were studied, but we considered principally the brain, skeletal muscles, myocardium, and kidneys. The main conclusion drawn from our results is that there is undoubtedly a direct correlation between tissue-water metabolism and tissue-energy metabolism.

Determination of Freezing Point of

Tissue Water by Infrared Spectroscopy

Determination of the freezing point of tissue fluids and solutions under different physicochemical conditions is one technique for studying the freezing and chilling of tissues and intact organisms; it has been specifically employed in the work of Louis Re [9] and Smith [10]. The pressing nature of these problems scarcely requires discussion.

Our laboratory has proposed a very simple method for determining the freezing point of a tissue from its infrared spectrum.

The $\nu_2 + \nu_L$ absorption band (2130 cm^{-1}) of water is recorded from a specimen placed in a special controlled-climate chamber with a smoothly falling temperature. The specimen temperature and the position of the absorption-band maximum are simultaneously registered. At the instant when the water in the specimen freezes, there is an abrupt change in the position of the absorption maximum from 2130 to 2240 cm^{-1}, to the position characteristic of ice. These water and ice bands are so close together and in such a convenient region of the spectrum that, by selecting appropriate experimental conditions, one can easily measure the freezing point of the tissue fluid with rather high accuracy. On the basis of average data from a large number of experiments, we estimate the error in our results to be 0.1°.

Figure 1 shows the change in the spectrum during the liquid—ice transition. The freezing point we obtained for tissues from the summer grass frog *Rana temporaria* are given below. The freezing temperature of the blood agrees with that given in the literature (—0.45°).

Tissue	Temp., °C
H$_2$O	—0.0
Blood	—0.5
Muscle	—0.7
Brain	—1.3
Kidney	—1.9

The fact that the freezing point of the tissue fluid in the brain was 0.6° lower than that in the muscles conforms to theories holding that water is more highly ordered in the brain than in muscle. The similar effect in the kidneys must apparently be attributed to binding of the water as a result of ionic hydration, the ion content of the kidneys substantially exceeding that of other tissues.

Thermal Changes in Near-Infrared Spectrum

of Water in Chick-Egg Albumin and Yolk

We also made a systematic study of the thermal changes in the 2130, 5200, and 6900 cm^{-1} bands in the spectrum of the water in chick eggs. The albumin and yolk were heated to 60° in one experiment and cooled to —10° under identical conditions in the other experiment, using a thermostated cell. The experimental results were compared with the thermal changes in the spectrum of ordinary liquid water, measured in the same apparatus. The water spectrum we obtained agreed with recent data in the literature on the temperature-related shifts in the near-infrared region of the water spectrum.

Similar data were published independently and almost simultaneously in articles by Buijs and Choppin [11], Yamatera, Fitzpatrick, and Gordon [12], and Luck [13]. These authors believe that the pronounced structure of the absorption bands and the trend of their individual maxima with temperature make it possible to draw quite definite quantitative conclusions regarding the existence of three types of molecules in liquid water. The S$_0$ molecules are not bound to adjacent molecules by hydrogen bonds (monomers). These molecules are especially large in number near the boiling point, the 5220 cm^{-1} absorption maximum corresponding to them. The S$_1$ molecules are bound to adjacent molecules by only one hydrogen bond, involving one hydrogen atom. These molecules are especially profuse near the freezing point and are characterized by the absorption band with a spectral frequency of 5160 cm^{-1}. Finally, the S$_2$ molecules are bound by two hydrogen bonds involving both hydrogens. They form a three-dimensional network and are predominant in ice, their presence being manifested in the band with a frequency of 5030 cm^{-1}.

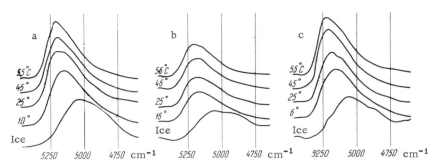

Fig. 2. Thermal changes in contour of absorption band of pure water (a),
egg yolk (b), and egg albumin (c).

Table 2. Proportions of Water Molecules of S_0, S_1, and S_2 Types (in %)

Freq., cm^{-1}	Water			Albumin			Yolk		
	Ice	Room temp.	55°	Ice	Room temp.	55°	Ice	Room temp.	55°
5210	16	37	41	18	37	41	19	36	40
5150	34	37	36	34	37	36	35	37	37
5050	50	26	23	48	26	23	46	27	23

Analogous absorption maxima appeared in the spectra we obtained for egg albumin and yolk. Their trend with temperature enables us to draw qualitative and even semiquantitative conclusions regarding the presence of water bound in different ways in eggs (Fig. 2).

We treated the temperature-related changes in the contour of the $\nu_2 + \nu_3$ band graphically, on the basis of elementary considerations following from examination of Fig. 2. Assuming that, in first approximation, the relative proportions of water molecules of the S_0, S_1, and S_2 types should be proportional to the intensities of the bands at the absorption maxima, we measured the amplitude of each band at three frequencies: 5210 cm^{-1}, i.e., the position of the high-frequency absorption maximum established at high temperatures ("near the boiling point" in Yamatera's terminology [12]); 5150 cm^{-1}, i.e., the position of the moderate maximum clearly observed at moderate temperatures in our graph ("near the melting point" according to Yamatera); and, finally, 5050 cm^{-1}, i.e., at the position of the ice absorption band. Table 2 presents the percentage proportions of the three types of molecules thus calculated for pure water, egg albumin, and egg yolk.

By virtue of its simplicity and demonstrativeness, we believe it possible to use this method for graphic treatment of spectral curves as a starting point for comparison of the spectra of biological tissues. Judging from the data in the literature, it is still impossible to find strict theoretical grounds for calculating the proportions of water molecules bound to different degrees [14], but preference among the approximations must indubitably be given to those that do not require lengthy calculation.

Our data for eggs differ so little from those obtained for pure water that we are forced to conclude that the water in eggs is identical in structural properties to ordinary water. It freezes at the same temperature as pure water and has the same solvent action. This conforms to the widely held view that most of the water in protein solutions is in the free state.

Investigation of Near-Infrared Spectra of

Alcohol – Water Solutions. Influence of Urea

on Solubility of Hydrocarbons in Water

The diverse effects of urea on biological processes, particularly its weakening of the hydrophobic bonds in protein and nucleic acid molecules, have now been related to its action on the structure of water.

Hydrocarbons that are insoluble or poorly soluble in water readily dissolve in aqueous urea solutions. An attempt was recently made to attribute this phenomenon to the ability of urea and water to form a joint cluster of requisite size around the hydrocarbon molecule. No increase in solubility under the action of urea is observed for hydrocarbons of small size (methane and ethane). This is due to the fact that these molecules fit rather conveniently into the voids of the open water structure and water—urea clusters are thermodynamically disadvantageous, since they are too open [15].

It is known that water-soluble nonpolar compounds having molecules of small size shift the structural equilibrium toward formation of aggregates that are ice-like or of the clathrate type. These compounds include the aliphatic hydrocarbons. Their solubility in water is extremely low and, in all probability, has an interstitial character, i.e., a solute molecule enters an existing void in the solvent structure, provided that the latter has dimensions suitable for the molecule in question. Solution of the compound is greatly impeded if this condition is not satisfied, since rearrangement of the solvent structure is necessary for solution to occur. When molecules of a nonpolar compound enter the voids in the ice-like (in the sense of form but not stability) structure, the latter undergoes some strengthening, which apparently depends to a large extent on the correspondence between the size and shape of the voids and those of the molecules introduced into them. The proportion of "structured" component in the mixture of open and densely packed water increases. It is possible to influence the solubility of nonpolar compounds by varying the size of the structural voids in some fashion. This model can be used to explain the decreasing solubility of hydrocarbons with temperature.

It is possible to influence the size of the cluster voids by solution of a compound that interacts strongly with water. Urea is a specific example of such a compound. It is known to have an unusually high solubility in water, reaching 20 M at 25°. There are data indicating that aqueous urea solutions are close to ideal at all concentrations. The interaction of water and urea molecules is so strong that it surpasses the interaction of water molecules with one another. Urea can compete with water for hydrogen bonds; a bond formed between urea and water molecules is stronger than a bond in pure water. The mechanism by which urea dissolves apparently differs radically from that by which hydrocarbons dissolve. Forming hydrogen bonds with water molecules, this compound produces composite clusters of both water and urea molecules. Moreover, despite the fact that the NH_2 group fits well into the open water structure because of the similarity between the H_2O and NH_2 molecules, the triangular shape of the flat urea molecule does not correspond geometrically to the tetrahedral arrangement of water. This creates difficulties in attempts to devise steric models of such clusters. Some of the water molecules are evidently displaced from the water structure and replaced by urea molecules, the size and shape of the structural voids being altered. It is most probable that the voids increase in size. If we compare this with the fact that urea in aqueous solutions increases the solubility of hydrocarbons having molecules of comparatively large size (propane and butane), it can be seen that this model fully explains the experimental data.

We determined the spectra of aqueous urea solutions at different concentrations and temperatures in the vicinities of the four water-absorption bands whose frequencies are composed

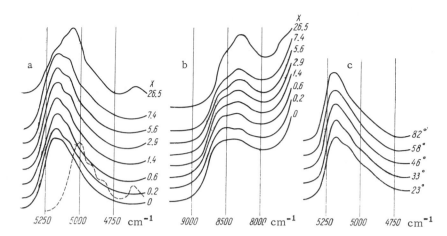

Fig. 3. Absorption spectra of aqueous urea solutions. a,b) Spectra at room temperature, urea concentration in mol.% indicated on graph; c) shape of 5200 cm^{-1} band as a function of temperature (in °C) at concentration of 0.2 mol.%. The broken line represents the spectrum of a urea emulsion in nujol.

of the fundamental-vibration frequencies (see Table 1). We consider the results obtained to confirm the existence of composite water−urea clusters, since solution of urea produced a new low-frequency maximum near the water-absorption band; the intensity of this band increased with the urea concentration and decreased with rising temperature.

This phenomenon was especially pronounced in the $\nu_2 + \nu_3$ band (5200 cm^{-1}), where an additional maximum developed at 5120 cm^{-1}; a maximum simultaneously appeared at 8350 cm^{-1}, near the $\nu_1 + \nu_2 + \nu_3$ band (8500 cm^{-1}) (Fig. 3). The total valence band $\nu_1 + \nu_3$ (6900 cm^{-1}) and the deformation-liberation band $\nu_2 + \nu_L$ (2130 cm^{-1}) were displaced in opposite directions (to 6850 and 2160 cm^{-1}, respectively), which indicates that hydrogen bonds stronger than in pure water were formed in both cases.

Urea absorption was observed in the form of a small maximum on the long-wave slope of the band at 5000 cm^{-1} in our spectrograms. For purposes of comparison, we determined the absorption spectrum of a urea emulsion in nujol, which is represented by the broken line in the graph.

We also investigated aqueous solutions of aliphatic alcohol (which enabled us to avoid the difficulties associated with the low solubility of aliphatic hydrocarbons in water). Alcohols are of interest because they have nonpolar alkyl groups and are also highly miscible with water. The presence of the OH group in the alcohol molecule can naturally cause material complication of the results. The alcohol molecule is dual in nature and probably occupies a position intermediate between the two extreme cases described above with respect to the mechanism of its solution in water [16]. On the one hand, the nonpolar alkyl groups enter the structural voids and, on the other hand, the OH group participates in forming the structure itself. It is probable that some rearrangement of the water clusters occurs during solution of an alcohol (as a result of the OH group), but the character of the interstitial solution process most likely remains unchanged. Most of the alcohol molecule enters a void in the water structure, while the hydroxyl "tail" is held in its wall. We can therefore expect some restriction of alcohol− molecule mobility. Actually, the literature contains data indicating that the rotary mobility of C_2H_5OH is restricted in aqueous solution [16].

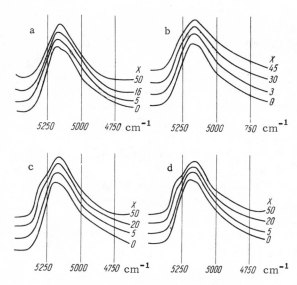

Fig. 4. Absorption spectra of aqueous alcohol solutions. a) Methanol; b) ethanol; c) propanol; d) tertiary butanol. The molar alcohol concentration is indicated in the graph.

The present paper presents the results of a study of the absorption spectra of alcohol-water solutions (water band) in the near-infrared region. It was important to select an area of the spectrum where the alcohol and water bands were not superimposed. The $\nu_2 + \nu_3$ water band located at 1.9 μ was naturally suitable from the experimental standpoint. None of the alcohols investigated had an absorption band in the immediate vicinity of this band. Our work was conducted in a spectrometer with a grating having a line spacing of 5-7 cm^{-1}.

We investigated methanol, ethanol, propanol, and tertiary butanol at mole-fraction concentrations between 0.01 and 0.50. Solutions of n-butanol were studied only at low water concentrations, because of the limited solubility of this compound.

In all the solutions investigated, the 5200 cm^{-1} water band shifted in the long-wave direction as the alcohol concentration increased (Fig. 4). This shift was largest in the methanol solution (30 cm^{-1}), smaller in the ethanol solution (18-25 cm^{-1}), and still smaller in the propanol solution (about 15 cm^{-1}). The shift in the tertiary butanol solution was about 23 cm^{-1}. The 5200 cm^{-1} band exhibited no shift at the n-butanol concentrations studied.

In addition to the shift, a shoulder appeared on the short-wave side of the band (at approximately 5300 cm^{-1}) as the alcohol concentration rose. The shoulder became a separate band in the spectrum of the n-butanol solution (the water concentration was very low in this case). This band belonged to unbonded, monomeric water molecules and was most easily seen in solutions of higher homologs.

In our opinion, this long-wave shift is a manifestation of stabilization of the water structure in aqueous alcohol solutions in comparison with pure water [17]. The decrease in the shift as the molecular weight of the alcohol increases is in conformity with the aforementioned theory of the interstitial solubility of alcohols in water.

The effect of tertiary butanol on the water spectrum is greater than might be expected from its molecular weight and its solubility in water is also high. This is due to the fact that it contains methyl groups, which fit well into the water structure.

The appearance of a monomeric band in the spectrum indicates that, when an alcohol is dissolved in water, the structural ordering of the solution as a whole is not increased but there is rather some sort of structural differentiation, which results in segregation of a region of water molecules bound to alcohol molecules and of monomeric water molecules not incorporated into this structure [18].

The authors wish to thank Professor M. F. Vuks for his concern with and interest in this work.

LITERATURE CITED

1. R. F. Bader and G. A. Jones, Canad. J. Chem., 41:586 (1963).
2. J. G. Bayly, V. B. Kartha, and W. H. Stevens, Infrared Physics, 3:211 (1963).
3. I. E. Bertie and E. Whalley, J. Chem. Phys., 40:1637 (1964).
4. O. Ya. Samoilov, Structure of Aqueous Electrolyte Solutions and the Hydration of Ions, Consultants Bureau, New York (1965); Zh. Strukt. Khim., 6:65 (798) (1965); H. S. Frank, Federat. Proc., Vol. 24, 2, Pt. III, March–April, p. 1 (1965); Ann. N. Y. Acad. Sci., Vol. 125, Art. 2, 730 (1965); Yu. V. Gurikov, Zh. Strukt. Khim., 6:66, 817 (1965).
5. N. A. Verzhbinskaya and A. I. Sidorova, Biofizika, 9:349 (1964).
6. N. A. Verzhbinskaya, in: Histohematological Barriers [in Russian], Izd. AN SSSR (1961), p. 146.
7. G. Herzberg, Vibration and Rotation Spectra of Monatomic Molecules [Russian translation], IL, Moscow (1949); Landolt-Boernstein, Zahlenwerte und Funktionen, Vol. 1, part 2, Molekeln, Berlin (1951), p. 333.
8. A. I. Sidorova and N. A. Verzhbinskaya, Zh. Prikl. Spektr., 3:525 (1965); Problems of Histohematological Barriers [in Russian], Izd. Nauka, Moscow (1965), p. 152; Biofizika, 11:101 (1966).
9. L. Re, Preservation of Life by Chilling [Russian translation], Izd. Med. Lit., Moscow (1962).
10. O. Smith, Biological Action of Freezing and Chilling [Russian translation], IL, Moscow (1963).
11. K. Buijs and G. R. Choppin, J. Chem. Phys., 39:2035 (1963); 40:3120 (1964).
12. H. Yamatera, B. Fitzpatrick, and G. Gordon, J. Mol. Spectrogr., 14:268 (1964).
13. W. Luck, Ber. Bunsenges. Phys. Chem., 67:186 (1963).
14. V. Vand and W. A. Senior, J. Chem. Phys., 43:1869 (1965); G. E. Walrafen, J. Chem. Phys., 40:3249 (1964).
15. M. Abu-Hamdiyyah, J. Phys. Chem., 69:2720 (1965); D. B. Wetlaufer, J. Am. Chem. Soc., 86:509 (1964).
16. F. Francs, Ann. N. Y. Acad. Sci., Vol. 125, Art. 2, 277 (1965).
17. M. N. Buslaeva and O. Ya. Samoilov, Zh. Strukt. Khim., 4:682 (1963).
18. A. I. Altsybeev, V. P. Belousov, and A. G. Morachevskii, in: Chemistry and Thermodynamics of Solutions [in Russian], Izd. LGU (1964).

CALORIMETRIC INVESTIGATION OF MACROMOLECULAR HYDRATION

G. M. Mrevlishvili and P. L. Privalov

Institute of Physics
Academy of Sciences of the Georgian SSR

Although the problem of the state of water near macromolecules has been intensively studied for several decades, using the entire arsenal of physical and chemical research methods, we still cannot say that we have progressed very far toward solving it.

One reason for this is the technical complexity of experimental research on the state of water and the fact that, in most cases, study of certain properties of water gives somewhat one-sided information, which is not easily related to other properties for want of a sufficiently well-developed theoretical model of water.

A second reason is the fact that the effect of a macromolecule on water is very complex and produces diverse changes in its state in a small volume around the macromolecule. It therefore seems improbable that the action of a macromolecule on water can, in general, be sufficiently well described by any one parameter.

Nevertheless, such parameters as hydration are often employed to characterize the macromolecule−water interaction. The term "hydration" refers to the amount of water under the influence (rather strong) of the macromolecule. This "influence" is usually described in terms of mobility, and water of hydration is considered to be water with an altered mobility or "bound" water, provided that the change in mobility is positive, which is not always the case.

The mobility of water molecules can theoretically be evaluated from NMR data, the dispersion of the dielectric constant, self-diffusion, or the activity coefficient. However, none of the aforementioned methods have made it possible to determine the hydration of macromolecules with a precision sufficient that the figures obtained can serve as a quantitative criterion.

A second approach to determination of hydration reduces to evaluating the H bonds in the water, which actually govern the water structure and figure directly in calculations of the factors determining the stability of the macromolecular conformation. Ordering of water near macromolecules should lead to an increase in the number of H bonds, bringing it close to the maximum level, which occurs in the ice structure. Bound water can be regarded as ice-like in this sense.

Data on the change in the number of H bonds near macromolecules can be obtained from infrared spectra, the chemical shift in the NMR lines, and analysis of thermochemical solution characteristics, but such evaluations are not yet common because of the complexity of interpreting the experimental data.

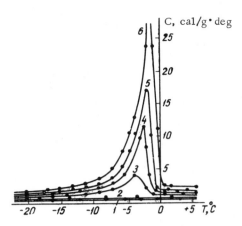

Fig. 1. Heat capacity of 1 g of procollagen as a function of temperature in the presence of different amounts of water: 1) 0 g; 2) 0.35 g; 3) 0.64 g; 4) 0.93 g; 5) 1.26 g; 6) 2.0 g.

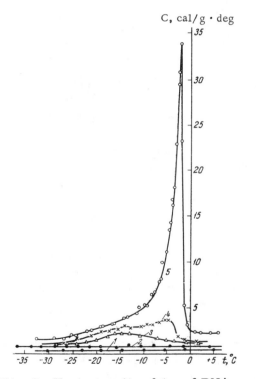

Fig. 2. Heat capacity of 1 g of DNA as a function of temperature in the presence of different amounts of water: 1) 0 g; 2) 0.5 g; 3) 0.75 g; 4) 1.0 g; 5) 2.0 g.

However, the effect of macromolecules on water can be characterized not only by the number of H bonds stabilized or ruptured, but also by the number of "frozen" or "thawed" water molecules, taking into consideration the fact that thawing is accompanied by breakage of a definite number of H bonds (judging from infrared data, about 1 mole of bonds per mole of molecules, i.e., about 50% of all the bonds, are broken during the melting of ice). By virtue of its graphic character, this presents definite advantages and points the way to direct experimental determination of bound water, using this term to mean water that does not undergo freezing or chilling and thus does not thaw when subsequently heated. This definition of bound water can be taken as basic (it is actually one of the first definitions of water in the bound state) and then no longer depends on initial premises regarding the variation in H bonding in water. We can move over to the latter by merely utilizing additional information on the relationship between state and number of bonds. If we take into consideration the fact that the mobility of water molecules is minimal in an ice-like structure with the maximum number of bonds, we can easily see the relationship between this definition of hydration and the dynamic definition in terms of mobility.

The amount of frozen water can be determined experimentally by various techniques, but, if we take into account the enormous specific heat of fusion of ice, it is obviously convenient to determine the proportion of melted water in the total amount of water in the system from the heat of fusion.

The advantages and demonstrativeness of the calorimetric method will be obvious from an examination of the following examples.

As can be seen from Fig. 1, which shows the heat capacity of 1 g of procollagen as a function of temperature in the presence of different amounts of water, the heat capacity of the dry procollagen preparation (curve 1) was a linear function of temperature over the entire range investigated. Addition of 0.3 g of water did not cause any material change in this function and no additional heat absorption was observed in the region where thawing of free water might be expected. The system therefore contained no free water. Additional heat absorption occurred only in the presence of 0.6 g of water per g of dry procollagen. This undoubtedly indicates that the system contained a small amount of water that underwent a

Table 1. Hydration of Macromolecules in Native and Denatured States

Preparation	Hydration, native molecules, g of H_2O/gram of dry weight, ± 0.005	Thickness of hydrate layer, Å	Hydration, denatured molecules, g of H_2O per gram of dry wt., ± 0.005
DNA	0.610	4.0	0.645
Procollagen	0.465	2.8	0.519
Serum albumin	0.315	3.2	0.330
Egg albumin.	0.323	3.4	0.332
Hemoglobin	0.324	3.8	0.339

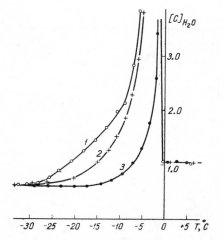

Fig. 3. Partial heat capacity of water as a function of temperature in macromolecular solutions containing 50%: 1) DNA; 2) procollagen; 3) hemoglobin.

structural transformation during chilling and was therefore not in the bound state. When the water content was further increased, the observed rise in heat absorption developed into a peak, whose maximum was shifted toward 0°.

A similar pattern was observed for a DNA preparation. As can be seen from Fig. 2, addition of water did not lead to any heat absorption in the vicinity of 0°C until a definite limit was reached. All the water added was therefore converted to a state that did not freeze on chilling and did not thaw on subsequent heating. In other words, it was in the bound state. Only when water was added in amounts of more than 0.5 g per gram of dry DNA did a rise in heat absorption occur, indicating that the water had H bonds that could be frozen and thawed by changes in temperature.

It is noteworthy that the trend of the water fusion curve depends directly on the nature of the macromolecule. As can be seen from Fig. 3, heat absorption extends over a substantially narrower temperature range for globular proteins than for the fibrillar procollagen; melting begins at still lower temperatures in the DNA solutions. This effect is apparently caused by the extensive contacts between the macromolecules and the water, the charge concentrations at the surface of the macromolecules, and the magnitude of the individual charges.

The heat of fusion or the observed peak area can be used to calculate the amount of water in a solution that melts when heated and thus the amount of bound water, as well as to evaluate the hydration of the macromolecules.

Table 1 presents the results of this type of calculation for different macromolecules in the native and denatured states.

The third column shows the thickness of the hydrate layer, calculated from the observed hydration and the geometric parameters of the native macromolecules, assuming that the hydrate layer is compact and uniformly covers the entire macromolecular surface (which is the only assumption that gives a rough figure for a simplified model of the structured water near macromolecules). The thickness of the hydrate shell is of the order of 1 or 2 molecular layers.

The next column gives the hydration of the denatured macromolecules. As can be seen, the hydration increases in all cases, although only slightly.

It must be noted that, because of its high sensitivity, the calorimetric method is as yet the only one that can detect this small difference between the hydrations of native and denatured macromolecules sufficiently clearly and unambiguously resolve the dispute that has been going on for years about the denaturational effect of hydration.

In addition to stating that the calorimetric method is actually very sensitive and merits the attention of researchers concerned with macromolecular hydration, we can draw the following conclusions from the data cited.

First, although the hydration of macromolecules is actually quite substantial, it is not as great as might be expected on the basis of Klotz's model of the macromolecule—water interaction.

Second, the observed denaturation effect confirms Kauzmann's model, which holds that the ordering effect of a macromolecule on water is increased by denaturation.

CALORIMETRIC INVESTIGATION
OF THE STATE OF TISSUE WATER

E. L. Andronikashvili, G. M. Mrevlishvili, and P. L. Privalov

Institute of Physics,
Academy of Sciences of the Georgian SSR

The problem of water in biological systems has been discussed for a number of years. The principal reason for this discussion is the fact that water is the main component of such systems, but nothing is as yet known about the state in which it is present or the role it plays.

The unusual properties of water and its ability to undergo ordering and to alter the conformation of macromolecules long ago gave rise to the belief that its role in biological systems is not merely that of a simple solvent, i.e., a medium in which macromolecules function by themselves. However, such statements have still not gone beyond the level of hypotheses and premises, since biological systems, by virtue of their complexity, have proved to be almost "opaque" to most of the physical and chemical techniques by which researchers have attempted to learn something about the state of water.

Assuming that the thermodynamic data on water in biological systems were both extremely exhaustive and reliable, we tried to resolve the problem from this direction. We had previously mentioned the possibility of this approach at a symposium on the biophysics of muscle contraction in Moscow in December, 1964.

Actually, the basic idea underlying the proposed approach is that if the water in biological systems is in some different, more ordered state than ordinary water, its partial absolute entropy should accordingly be less than the absolute entropy of ordinary "spring" water. This notion, which follows directly from thermodynamic principles, is so obvious that it requires no explanation and, at first glance, can only cause us to wonder why no one had previously utilized it, preferring instead to determine the state of water by various indirect and very complicated methods.

However, theoretical feasibility does not mean practical feasibility and closer examination shows that determination of the partial absolute entropy of water in biological systems is in fact not so simple and, moreover, is apparently impossible to accomplish by ordinary physical techniques such as the calorimetric method, since we cannot see how it could be pos-

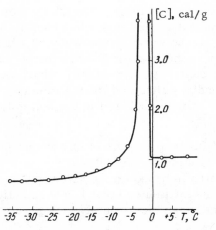

Fig. 1. Partial heat capacity of water in frog muscle as a function of temperature.

Table 1. Energy Parameters for Certain Tissues

Tissue	Partial specific enthalpy of fusion, H°, cal	Specific heat of fusion of ice, H°, cal	Partial specific entropy of fusion, S, e.u.	Specific entropy of fusion of ice, S°, e.u.	Entropy differ. per gram dry wt., ΔS, e.u.	Structural temp. of water in tissue, ΔT°	Amt. bound water per gram of dry wt., mg
Rat brain	75.96	79.70	0.2793	0.2919	—0.0445	—3.4	0.15
Frog muscle	75.44	—	0.2773	—	—0.0703	—4.0	0.25
Frog liver	70.46	—	0.2590	—	—0.1192	—8.9	0.41
Silk-worm eggs	71.0	—	0.2627	—	—0.1110	—8.6	0.38

sible to determine the residual entropy of a system at absolute zero.

Nevertheless, it turns out that the state of water in biological systems can be evaluated by this method, since only that portion of the entropy that goes to change the aggregate state is actually important in this case. It can be determined experimentally by measuring the heat absorption during thawing of the system.

Figure 1 shows a typical example of the heat absorption observed during thawing of a frog muscle. As can be seen, heat absorption began at —22° and continued until —0.5° was reached.

The extended character of the thawing processes creates definite difficulties in isolating the portion of the heat that actually goes into rearrangement of the water structure, i.e., melting. However, by using a specially devised calculation technique, which is based on linear extrapolation of the heat capacity and takes into account the accumulation of the thawed phase, we can determine the heat (enthalpy) and entropy of thawing with rather high accuracy.

The values thus found for various tissues are summarized in Table 1. As can be seen, the enthalpy and entropy of thawing of a gram of water are far less than the enthalpy and entropy of fusion of ice. This difference is very substantial for the entropy calculated per gram of dry weight.

The observed entropy deficit can be interpreted as meaning that the ordering of the water in the tissues investigated is greater than the ordering of ordinary water. It can take either of two forms: 1) ordering of the water structure as a whole (in which case it is convenient to characterize it in terms of the "structural temperature," i.e., the water temperature at which the entropy coincides with the observed entropy of the water in the tissue); 2) ordering of part of the water adjoining the intracellular structures, in which case it is convenient to characterize it by the amount of bound water. The latter factor is readily calculated, taking into account the entropy of fusion of pure ice. The values obtained are summarized in the last column of the table.

Judging from the fact that the deficit in the entropy of fusion per gram of dry weight in macromolecular solutions (see p. 65) is independent of the dilution once a critical concentration has been reached, we believe that the second model, which is based on the heterogeneity of the state of water, corresponds more closely to the actual situation and that it makes better sense to evaluate the amount of bound water in a tissue rather than the structural temperature.

As can be seen from the figures given, the amount of bound water in tissues is not very large, in some cases being even less than that present in protein solutions. This apparently results from the fact that the molecules in tissue form submolecular structures and the number of contacts between the molecules and the water is therefore substantially less than in solution. Hence it follows that the values obtained for the amount of bound water can be used to evaluate the total contact surface between the nonaqueous cellular structures and the water, or the "dispersion" of these structures.

HYDROPHOBIC BONDS IN PEPSIN

A. A. Vazina and V. V. Lednev

Institute of Biological Physics
Academy of Sciences of the USSR

Introduction

This article discusses the problem of hydrophobic interactions in pepsin, utilizing data from the literature and our own experimental results.

The concept of "hydrophobic bonds" (interactions) is more and more frequently called upon to explain the physicochemical properties of macromolecules. There are a few works, principally theoretical or qualitatively experimental, devoted to hydrophobic interactions in proteins. The theory of hydrophobic interactions is most extensively covered in the articles of Scheraga and Nemethy [1-3], who presented a statistical treatment of hydrophobic forces based on the models of structured water devised by Bernal [4-6] and Frank [7]. A hydrophobic interaction in a protein is understood to refer to the interaction of nonpolar residues of such amino acids as alanine, valine, leucine, cystine, methionine, phenylalanine, etc. According to Scheraga and Nemethy [2], the stability of such bonds or (in thermodynamic terms) the negative free energy of their formation results from a decrease in the ice-like character of water during formation of hydrophobic bonds. As is well known [8], dissolution of hydrocarbons (used to simulate nonpolar amino acids) is accompanied by a decrease in specific entropy of about 20 entropy units per mole and only a slight change in enthalpy (0-1 kcal/mole). The enthalpic constituent is responsible for the negative change in free energy during formation of hydrophobic bonds. The increase in entropy can be explained in the following manner: water is subject to continuous formation and melting of regions with an ordered structure similar to that in ice, in which the water molecules retain a roughly tetrahedral coordination produced by four directed intermolecular hydrogen bonds. However, severe stretching and breakage of the hydrogen bonds takes place at the same time, so that there are water molecules with from 0 to 4 hydrogen bonds. The relative number of water molecules with a given number of hydrogen bonds follows a Boltzmann distribution ($N \sim e^{-E/kT}$, where E is the energy of the molecule). Molecules with 3, 2, 1, or 0 hydrogen bonds can participate in interactions of the dipole-induced dipole type with nonpolar amino acid residues, their energy increasing in this case. According to the Boltzmann formula, the molecular distribution then shifts toward an increase in the number of molecules with saturated hydrogen bonds, i.e., the structuring (ice-likeness) of the water increases. It is understandable that the ice-likeness of the solution becomes greater with the contact surface between the water molecules and the nonpolar residues; conversely, the ordering of the solution decreases when the nonpolar residues aggregate, which leads to a negative change in free energy and is manifested in stability of such aggregated nonpolar segments of a protein. As Scheraga noted, aggregation may not be important for small hydrocarbon molecules, since the rotational degrees of freedom are lost in this case. However, such aggregation is more important in proteins, since the backbone of the polypeptide chain is to a large extent rigidly fixed.

Calculations have shown that the energies of hydrophobic and hydrogen bonds are comparable. The influence of hydrophobic bonds on the kinetic and equilibrium properties of proteins has been demonstrated and qualitatively evaluated in articles by Scheraga [9] and Scheraga and Nemethy [1-3]. In particular, these authors showed that hydrophobic bonds, in contrast to hydrogen bonds, do not weaken with rising temperature until a definite limit (65°) is reached and may even be somewhat strengthened. This also follows from the fact that solution of hydrocarbons (which we used as a model of hydrophobic bonds) is an exothermic process.

Experimental investigation of the influence of hydrophobic bonds on the kinetic and equilibrium properties of proteins is hampered by the complexity of segregating the effect of hydrophobic interactions in pure form. For example, the effect of the hydrophobic interactions in hemoglobin (whose existence was proved by determination of the structure of this compound [10]) are masked by the effect of the hydrogen bonds in the α-helical segments of the molecule.

Hydrophobic Interactions in Pepsin

Hydrophobic interactions apparently occur in purer form in pepsin than in other proteins. The present paper is an attempt to generalize the available data on the hydrophobic interactions in this compound.

Various investigations of pepsin have shown that this proteolytic enzyme has unusual physicochemical properties. It has a molecular weight of 34,500 and evidently consists of a single polypeptide chain containing 341-345 amino acid residues [11]. Of these, 60% are nonpolar (glycine, valine, leucine, isoleucine, alanine, methionine, etc.), 71 are dicarboxylic amino acids, 36 are free carboxylic residues, and only 4 are basic residues (one lysine, one histidine, and two arginines) [12]. The optical rotation and dispersion of optical rotation also show that pepsin does not have an α-helical chain configuration [12-13]. The optical rotation of pepsin is small, while the λ_C-constant of the dispersion of optical rotation is 216 (a typical figure for denatured proteins) and is not altered by exposure to urea in concentrations of up to 4 M. Enzyme activity persists in this case [12-14], i.e., hydrogen bonds apparently do not participate in stabilizing the globular configuration. The pepsin molecule is folded into a compact globule by the hydrophobic interaction of the nonpolar amino acid residues [3, 14, 15]. This hypothesis has also been confirmed by studies of the kinetics of thermal denaturation of pepsin in ethanol solutions of different concentrations [16].

Kinetics of Thermal Denaturation of Pepsin

Edelhoch investigated the kinetics of thermal denaturation of pepsin in alcohol solutions of different concentrations [16] and compared his results with those of a similar study of hemoglobin [17] (Table 1). As Edelhoch noted, the denaturation rate increases in both cases, but a rise in the alcohol concentration has opposite effects on the activation parameters of pepsin and hemoglobin. The decrease in $\Delta S*$ (the change in activation entropy) for pepsin and the increase in this factor for hemoglobin indicate that the molecules of the former pass into a less-ordered state, while those of the latter pass into a more-ordered state. The influence of alcohol on $\Delta H*$ and $\Delta S*$ (where $\Delta H*$ is the change in activation enthalpy) should depend on the relative contributions made by hydrogen and hydrophobic bonds. The differing character of the changes in $\Delta S*$ and $\Delta H*$ for pepsin and hemoglobin indicates that most of the amino acids in the hemoglobin molecule, which has a molecular weight of 34,000, are hydrogen-bonded in the native protein, while there are few hydrogen bonds in pepsin, which has a molecular weight of 34,500. It is obvious that saturation of the hydrogen bonds and weakening of the hydrophobic bonds occurs in hemoglobin when alcohol is added, while the pepsin molecule passes from a more-ordered into a less-ordered state as a result of breakage of the hydrophobic bonds.

Table 1. Influence of Ethanol on Kinetics of Thermal Denaturation
of Pepsin and Hemoglobin

pH	Temp. region, °C	Alcohol conc., %	ΔH^*	ΔF^*	ΔS^*
		Pepsin			
5.90; 0.15 -NaCl	41—49	0	95	226	23.0
	31—36	20	88	215	22.0
	21—29	40	52	104	21.2
		Hemoglobin			
6.00—7.00	—	0	76	153	25.1
	—	20	107	264	23.5
	—	40	117	309	22.5

Table 2. Structural Parameters of Pepsin Molecule

Protein modification	Radius of inertia, Å	Ratio of axes of equivalent ellipsoid of rotation	Axes of ellipsoid, Å	Vol., Å	Surface-to-volume ratio, Å^{-1}
Native pepsin at pH 5.6	20.5	2	37 74	55000	0.26
Pepsin + 40% ethanol (0.15-N NaCl)	28	3	38 114	133000	0.15

Edelhoch's results indicating that both hydrogen and hydrophobic bonds contribute to the conformation of pepsin and hemoglobin are in good agreement with the data yielded by x-ray diffraction analysis of the structure of hemoglobin and by physicochemical studies of pepsin.

Investigation of Structural Parameters

of Pepsin Molecule

We employed diffuse scattering of x rays at small angles [18] to investigate the change in the structural parameters of the pepsin molecule during heating and when the solvent polarity was changed (Table 2). Ethanol was used as a nonpolar solvent. All the structural parameters of the pepsin molecule were markedly altered at an ethanol concentration of 40%. The sharp jump in the radius of inertia in a 40% alcohol solution was due to weakening of the hydrophobic interactions in the pepsin. The molecular volume greatly increased and the surface-to-volume ratio dropped from 0.26 to 0.15 Å^{-1}. In order to demonstrate that the conformational transition does not result from electrostatic interactions (it will be remembered that pepsin contains 71 dicarboxylic amino acids), we also made measurements with pepsin solutions in which these interactions were suppressed with 0.15-N NaCl and obtained the same results.

Since the hydrophobic regions of the pepsin molecule contain tryptophan, tyrosine, and phenylalanine (6, 8, and 14 residues, respectively), a change in the environment of the chromophoric groups of these amino acids should affect the absorption of pepsin solutions in the vicinity of 280 mμ. The change in absorption at 280 mμ paralleled that in the radius of rotation. The optical density jumped from 0.700 to 0.780 at an alcohol concentration of 40%.

Comparison of our results with those obtained by Edelhoch shows that the structural changes in pepsin parallel those in the heat and entropy of activation, while the sudden jumps in the activation parameters of the molecule, i.e., its biological activity, are closely if not directly related to conformational transformations.

Thermal Stability of Pepsin

As was mentioned in the introduction, hydrophobic bonds, in contrast to hydrogen bonds, are not weakened at elevated temperatures and may even be somewhat strengthened. In this connection, we expected to find the conformation of the pepsin molecule to be thermostable if it were actually produced by hydrophobic bonds. We demonstrated in a previous article [18] that the optical density of a solution remains unchanged when the temperature is raised from the ambient level to 80°.

Moreover, study of the scattering of x rays at small angles by a 2% pepsin solution at 70° shows that the geometric parameters of the heated molecule are identical to those at room temperature. This apparently is still the only example of thermostability of the hydrophobic bonds of a protein.

Pepsin is known to be comparatively thermostable [19]. It loses only 25% of its enzyme activity in solution when held at 56° for 6 h. Dry pepsin can be heated to 100–120° without loss of enzyme activity. The enzymatic stability of pepsin apparently results from the thermostability of the conformation of the polypeptide chain, which is held in a compact globule by the hydrophobic interaction of the nonpolar amino acids.

In connection with the thermostability of pepsin, it is interesting to note that the pepsin predecessor pepsinogen can be heated to the boiling point and will regain its potential activity when cooled in the absence of salt.

Conclusions

1. We have considered the role of hydrophobic bonds in pepsin.

2. It has been demonstrated that various physicochemical studies of pepsin presuppose that its conformation is determined principally by hydrophobic bonds.

3. Pepsin has been used to confirm experimentally the hypothesis that the hydrophobic bonds in a protein are thermostable.

LITERATURE CITED

1. G. Nemethy and H. A. Scheraga, J. Chem. Phys., 36:3382, 3401 (1962).
2. H. Scheraga and G. Nemethy, J. Phys. Chem., 36:1773 (1962).
3. H. Scheraga, H. Nemethy, Z. Izchak, and R. J. Steinberg, J. Biol. Chem., 237:2506 (1962).
4. J. Bernal, Nature, 183:141 (1959).
5. J. Bernal, Proc. Roy. Inst. Gr. Brit., 37(4):355 (1959).
6. J. Bernal, Scient. Amer., Vol. 203, No. 2 (1960).
7. H. S. Frank and M. J. Evans, J. Chem. Phys., 13:507 (1945).
8. W. Kauzmann, Advances in Protein Chem., 14:1 (1959).
9. H. Scheraga, J. Phys. Chem., 65:1071 (1961).
10. M. F. Perutz, J. Mol. Biol., 13:646 (1965).
11. O. O. Blumenfeld and G. E. Perlan, J. Gen. Physiol., 42:553 (1959).
12. B. Jirgensons, Arch. Biochem. and Biophys., 74:70 (1958).
13. G. Perlman, Proc. Nat. Acad. Sci., 45:915 (1959).
14. J. Steinhardt, J. Biol. Chem., 123:543 (1938).
15. G. E. Perlman, Arch. Biochem. and Biophys., 65:210 (1955).

16. H. Edelhoch, Biophys. et Biochim. Acta, 38:113 (1960).
17. Booth, Biochem. J., 24:1699 (1930).
18. A. A. Vazina, V. V. Lednev, and B. K. Kemazhikhin, Biokhimiya, 31:720 (1966).
19. E. E. Ganassi, N. E. Kondakova, G. K. Otarova, and L. Kh. Eidus, Radiobiologiya, 1:14 (1961).
20. F. Putnam, in: Proteins, Vol. 2 (edited by H. Neurath and K. Bailey) [Russian translation], IL, Moscow (1953), p. 682. [English edition: Academic Press, New York (1954).]

ROLE OF EFFECTS PRODUCED BY
HYDRATION OF IONIC GROUPS
IN STABILIZING THE GLOBULAR STRUCTURE
OF CERTAIN PROTEINS

G. I. Likhtenshtein

Institute of Molecular Biology
Academy of Sciences of the USSR

Hydration is known to have a material effect on the energy of the hydrophobic, hydrogen, and electrostatic bonds that maintain the tertiary globular structure of proteins [1, 2]. On the basis of an analysis of data on denaturation kinetics, the amino acid composition of a number of proteins, and the thermodynamic properties of certain model compounds in aqueous solutions, the present paper attempts to establish which of the aforementioned interactions plays the principal role in stabilizing the native form of the proteins considered.

According to current theories, the processes by which biopolymers are denatured are conformational transitions that proceed by a cooperative mechanism [3-5]. Their cooperativity is manifested in the fact that, as a result of the interaction between individual elementary segments (elements) of the system, the probability that a segment will pass from one state into another depends on the state of the adjacent segments. The thermodynamic theory of cooperative systems has been elaborated in the literature [3, 4, 6-8]. Theoretical studies of the kinetics of cooperative processes have been devoted principally to analysis of the initial non-steady-state stages [9-13]. At the same time, it has been established that the denaturation of many proteins is described by simple rules. This enables us to assume that denaturation processes take place under quasi-steady-state conditions. A simple one-dimensional model is used below to analyze the basic kinetic mechanisms of quasi-steady-state cooperative transitions. We will subsequently consider the changes introduced into these mechanisms by consideration of the real properties of three-dimensional globular systems and compare the conclusions drawn from this theory with experimental data on protein denaturation.

Let us assume that a unit volume contains X_1 identical one-dimensional systems, each of which consists, in the general case, of N different elements n_i: $n_1 - n_2 - .. - n_{N-T} n_N$ (I). Each of the elements n_i can theoretically pass into the different state α_i (the transition $n_i - \alpha_i$ can specifically correspond to breakage of a certain bond); the ultimate result of the process is considered to be the appearance of the product $d_1 - d_2 - .. - d_{N-1} \cdot d_N$ (II). Let us also assume that, because of the strong cooperative bonding between adjacent elements, spontaneous transition of the elements occurs at a negligibly small rate in comparison with the rate at which some very weakly bonded element n_1 undergoes transition. Transition of n_1 initiates transition of n_2, the latter facilitates transition of n_3, etc. The process accordingly follows the schemes:

$$n_1 - n_2 - \ldots - n_{N-1} - n_N \underset{K_{-1}}{\overset{K_1}{\rightleftarrows}} d_1 - n_2 - \ldots n_{N-1} - n_N \underset{K_{-2}}{\overset{K_2}{\rightleftarrows}} d_1 d_2 - \ldots$$

$$- n_{N-1} - n_N \underset{K_{-3}}{\overset{K_3}{\rightleftarrows}} \cdots \underset{K_{-(N-1)}}{\overset{K_{N-1}}{\rightleftarrows}} d_1 - d_2 - \ldots d_{N-1} - n_N \underset{K_{-N}}{\overset{K_N}{\rightleftarrows}} d_1 - d_2 - \ldots - d_{N-1} - d_N,$$

where K_i and K_{-i} are the rate constants of the forward and reverse transitions of the i-th structural element.

If the elementary transformations of individual elements proceed at rates that exceed the rate of the first initiatory stage, a quasi-steady-state (with respect to intermediate substances) regime is established in the system.

The rate constant of the appearance of the product of reaction (II) can be found from Temkin's formula [14]:

$$K = K_1 \left[1 + \frac{K_{-1}}{K_2} + \frac{K_{-1}}{K_2} \frac{K_{-2}}{K_3} + \ldots + \frac{K_{-1} K_{-2} \cdots K_{-(N-1)}}{K_2 \cdot K_3 \cdots K_{N-1} K_N} \right]^{-1} = K_1 \, \nu, \tag{1}$$

where $\nu = K/K_1$ is the length of the cooperative chain, which equals the number of product molecules produced by a single act of initiation.*

In the special case where all the elements of the ensemble (except the first) are identical:

$$K_{-1} = K_{-2} = \ldots K_{-|N-1|} = K_- \text{ and } K_2 = K_3 = \ldots K_N = K_+$$

$$K = K_1 [1 + \overline{K} + \overline{K}^2 + \ldots \overline{K}^m]^{-1} = K_1 \frac{\overline{K} - 1}{\overline{K}^{m+1} - 1}, \tag{2}$$

where $m = N - 1$ and $\overline{K} = K_-/K_+$ is the equilibrium constant of the elementary transformation.

The experimental activation energies (E_{eff}) are found from the Arrhenius formula:

$$E_{eff} = RT^2 \, (d \ln K/dT) \cdot \tag{3}$$

Substituting (2) into (3) after simple transformations, we find that

$$E_{eff} = E_1 + \overline{K}/(\overline{K} - 1) \times \Delta H - \overline{K}^{m+1}_{(m+1)} \cdot \Delta H_0/(\overline{K}^{m+1} - 1), \tag{4}$$

where E_1 is the activation energy of initiation and ΔH_0 is the enthalpy of the elementary transition of a single structural element.

It follows from Eqs. (2) and (4) that:

1. At $K_- \gg K_+$, i.e., at $\overline{K} \gg 1$, the rate constant $K = K_1/\overline{K}^m$ is substantially less than the rate constant of initiation; in this case,

$$E_{eff} = E_1 + m \cdot \Delta H_0 \tag{5}$$

and

$$\Delta S^{\neq}_{eff} = \Delta S^{\neq}_1 + m \cdot \Delta S_0, \tag{6}$$

where $\Delta S^{\neq}_{eff}$ and $\Delta S^{\neq}_1$ are the experimental activation entropy and the activation entropy of the initiation stage, respectively, while ΔS_0 is the entropy of the elementary transition.

2. At $K_+ = K_-$, i.e., $\overline{K} \to 1$, the total rate constant begins to rise and approximates the value K_1/m.

---

*The formal analogy between chain and cooperative processes was pointed out in a previous article [13].

3. At $K_+ \gg K_-$, i.e., $K \ll 1$, the rate of the process is almost equal to the initiation rate, regardless of the number of intermediate stages: $K = K_1$, $E_{eff} = E_1$, and $\Delta S_1^{\neq} = \Delta S_{eff}^{\neq}$.

In the example under consideration, the quantity m in the expressions for the experimental K, E_{eff}, and $\Delta S_{eff}^{\neq}$, equals the number of participants in the cooperative ensemble. According to Eq. (1), if one of the intermediate stages becomes irreversible, this leads to disappearance of all the terms reflecting the contribution of all following stages, regardless of whether they are reversible or not. In the more general case, any stage can be limiting and the process can be described by Pshenichnov's formula [15]:

$$K = K_j \left[1 + \frac{K_{-j}}{K_{j+1}} + \ldots + \frac{K_{-j} \cdot K_{-(j+1)} \cdots K_{-(l-i)}}{K_{j+1} \cdot K_{j+2} \cdots K_e} \right]^{-1}, \tag{7}$$

where K_j is the rate constant of the j-th (limiting) stage and K_l is the rate constant of the first irreversible stage. It is obvious that $(l - i) \leq N - 1$.

In the case of two-dimensional and three-dimensional cooperative systems, it is possible for the processes to follow several linear routes and the expressions for the effective rate constants and activation energies can theoretically be found by the methods developed for complex steady-state reactions [14, 16, 17]. The values of K and E_{eff} will include combinations of sums and products, which will in turn depend on the specific sizes and shapes of the cooperative ensembles. It is obvious, however, that the maximum effective activation energy at $\overline{K} \gg 1$ will be governed by the term that is the product of the maximum equilibrium constant (m) for the system in question, which reflects the contribution made by the shortest reversible route. At $\overline{K} \ll 1$, the total rate of the process equals the rate of the initiating (or limiting) stage and $E_{eff} = E_i$, as in a one-dimensional system.

Initiation in real systems can begin with any structural element and at several places simultaneously. It is readily shown that these circumstances can lead only to a decrease in m and thus in E_{eff}. Heterogeneity of the system also causes a decrease in E_{eff} [18, 19].

Qualitative evaluation of the principal features distinguishing real cooperative systems of the globular-protein type from a simple one-dimensional model thus enables us to conclude that the experimentally determined activation energy of denaturation

$$E_{eff} \leqslant E_1 + m \cdot \Delta H_0 \leqslant E_1 + N \cdot \Delta H_0, \tag{8}$$

where N is the total number of elements participating in the transition (e.g., the total number of bonds broken during denaturation) and m characterizes the length of the shortest continuous cooperative segment in the system; this segment begins with the element whose transition limits the process rate and ends with a segment that undergoes an irreversible transformation. All the transitions within the segment are reversible.

Equation (8) permits us to use the experimental values of the E_{eff} of denaturation to estimate the lower limit of N.

An important property of cooperative processes is the fact that a difference in the value of m has the same effect on the values of E_{eff} and $\Delta S_{eff}^{\neq}$ in Eqs. (5) and (6). Substituting and canceling, these equations yield the relationship

$$E_{eff} = E_i - \frac{\Delta H_0}{\Delta S_0} \Delta S_i^{\neq} + \frac{\Delta H_0}{\Delta S_0} \cdot \Delta S_{eff}^{\neq} \cdot \tag{9}$$

At sufficiently large m, E_{eff} and $\Delta S_{eff}^{\neq}$ are far more sensitive to changes in the parameters ΔH_0, ΔS_0, and m than to corresponding changes in E_i and ΔS_i. The dependence of E_{eff} and

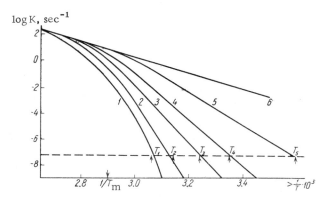

Fig. 1. Theoretical relationships between log K and 1/T for cooperative processes with different values of m. E_1 = 24 kcal, ΔS_1 = 15 eu, ΔH_0 = 2 kcal, S_0 = 2.8 eu, and m: 1) 200; 2) 140; 3) 80; 4) 60; 5) 30; 6) 0.

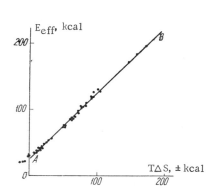

Fig. 2. Relationship between E_{eff} and $T_{tl} \cdot \Delta S_{eff}$ for protein denaturation [21–24], A—B being the theoretical line.

$\Delta S_{eff}^{\neq}$ should therefore be described by the equation*

$$E_{eff} = \Delta F_i^{\neq} + T_m \cdot \Delta S_{eff}^{\neq} , \tag{10}$$

where $T_m = \Delta H_0 / \Delta S_0$ is the "melting" point, at which $\bar{k} = 1$, and $\Delta F_i^{\neq} = \Delta H_i^{\neq} - T_m \cdot \Delta S_i^{\neq}$ is the free energy of activation of the initiating (limiting) stage at $T = T_m$.

Figure 1 shows the theoretical relationships between log K and 1/T for cooperative systems with different values of m. The horizontal broken line characterizes the sensitivity of the method used to measure K.

The process rate should rise from almost zero to its final level at certain temperatures T_i close to but not equal to T_m. As we approach the true melting point, the temperature coefficient K should decrease to the value of the temperature coefficient of initiation.

Figures 2 and 3 present a few examples illustrating the wide occurrence of compensation mechanisms in protein denaturation [22–25].

Equation (10) is quite precisely satisfied for many proteins, which is apparently an experimental confirmation of the cooperative mechanism of denaturational processes.

The experimental activation energies of thermal denaturation lie between 20 and 200 kcal, the activation entropies are between 0 and +600 eu, and the melting points (or, more precisely, T_i) are between 330 and 350°K. Bond breakage during denaturation is therefore accompanied by an increase in ΔH_0 and ΔS_0, while $\Delta H_0 / \Delta S_0 \approx$ 330–350°K. The activation energy decreases to 30–10 kcal during denaturation with acids and alkalis (Fig. 5); T_i drops to 270–310°K. It can be seen from Eq. (10) and Fig. 2 that the free energies of initiation $\Delta F_i^{\neq}$ for different proteins are similar and lie between −24 and 26 kcal.

Moistening of dry protein preparations leads to an increase of 30–60 kcal in E_{eff}, which indicates that zones of continuous cooperativity develop during protein hydration [13, 25, 26].

---

*Relationships similar to Eq. (10) have been obtained from more general premises in some articles [13, 20, 21].

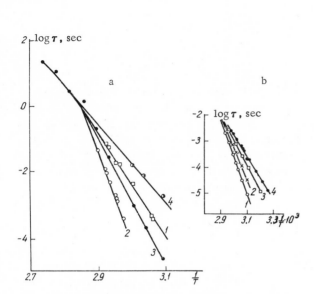

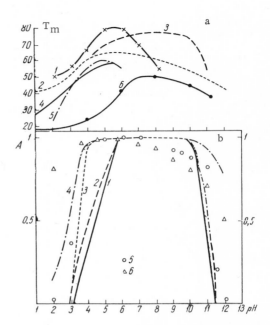

Fig. 3. Relationship between log τ and $1/T$ (τ is the time required for the initial native-protein concentration to decrease by a factor of e). a) Catalyse, bovine liver: 1) pH 7, H_2O [25]; 2) pH 7, D_2O [25]; 3) pH 7, stable fraction; 4) pH 7, labile fraction; functions 3 and 4 were obtained in conjunction with T.V. Troshkina and Yu. M. Azizov. b) bacterial α-amylase: 1) without additive; 2,3,4) urea concentrations of 1 M, 3 M, and 6 M, respectively [24].

Fig. 4. Influence of pH on stability of certain enzymes and proteins. a) T_m of proteins as a function of pH: 1) insulin film [30]; 2) insulin solution [30]; 3) egg albumin [32]; 4) ribonuclease solution [50]; 5) ribonuclease film [50]; 6) fibrin fibers [51]. b) Relative stability of enzymes as a function of pH: 1) bacterial α-amylase [47]; 2) pancreatic α-amylase [47]; 3) β-amylase [48]; 4) papain [49]; 5) catalyse, bovine liver; 6) peroxidase, horseradish (data for curves 5 and 6 were obtained in conjunction with T. V. Troshkina and Yu. M. Azizov).

Breakage of which type of bond is responsible for the experimentally observed kinetic phenomena in globular proteins ?

Hydrophobic bonds have recently been assigned a major role in maintaining the tertiary structure of proteins [7, 27-29]. However, it is known that breakage of these bonds in model systems is generally accompanied by a decrease in enthalpy and an even larger drop in entropy [1, 30], which should produce values of T_m below 200°K and reduce E_{eff} and $\Delta S_{eff}^{\neq}$ of denaturation. The value of ΔH_0 becomes positive on rupture of a hydrophobic bond only at temperatures above 60° [30]; it never exceeds +1 kcal/mole. Using this known elevated value and calculating the maximum number of hydrophobic bonds possible for a given protein, we can use Eq. (8) to estimate the maximum contribution that can be made by hydrophobic bonds to the activation energy of protein denaturation of ΔE_{max}.

It follows from Table 1, which compares the calculated values of ΔE_{max} with the experimentally determined activation energies of denaturation E_{eff}, that the contribution made by hydrophobic groups is clearly inadequate to account for the high experimental values of E_{eff} for a number of proteins, despite the fact that this quantity includes a factor known to be less than the number of broken bonds N. The stability of many proteins is greatly reduced in acidic

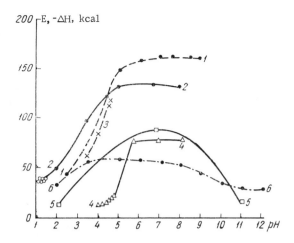

Fig. 5. Influence of pH on activation energy (E_{eff}) and heat of protein denaturation. 1) Egg albumin (ΔH) [32]; 2) egg albumin (E_{eff}) [22]; 3) human serum albumin [42]; 4) hemoglobin [22]; 5) catalyse, bovine liver; 6) peroxidase, horseradish (data for curves 5 and 6 were obtained in conjunction with T. V. Troshkina and Yu. M. Azizov).

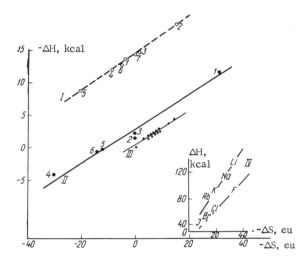

Fig. 6. Relationship between standard enthalpies (ΔH) and entropies (ΔS) of certain processes. I) Formation of Cu^{++}-dimethylene diamine complex in aqueous salt solutions: 1) without additives; 2) 0.1 N KCl; 3) 0.15 N $NaClO_4$; 5) 0.5 N KNO_3; 6) 1.0 N KNO_3; 7) 1.8 N KNO_3 [35]. II) Formation of ionic pairs in aqueous solutions: 1) $[CeClO_4^-]$; 2) $[MgCH_3COO^-]^+$; 3) $[MgCHCOO^-]^+$; 4) $[CeSO_4]^+$; 5) $[CaCH_3COO^-]^+$; 6) $[CaCHCOO^-]^+$ [35]. III) Difference in heats of solution of amino acids in water and water—alcohol solutions as a function of difference in entropies of solution (dots). The line corresponds to the analogous function with varying temperature. Calculated by the author from solubility data [36]. IV) Ionic hydration [34].

and alkaline media (Fig. 4); this drop in stability is accompanied by a sharp decrease in the E_{eff} of protein denaturation and enzyme deactivation (Fig. 5). It is also difficult to explain these phenomena within the framework of theories holding that hydrophobic bonds are the most important factor.

The character of the dependence of protein stability on pH indicates that a substantial contribution is made by the bond between the carboxylate ion and the ionized amino groups of lysine for the phenol group of tyrosine [30–33]. There are two theoretical possibilities: 1) the bond is embedded in the hydrophobic "drop" formed by the protein globule; 2) the bond is formed in a fold of the globule and is surrounded by water molecules oriented both by the electrostatic forces of the ions and by the adjacent hydrophobic groups (a similar situation apparently obtains in hydrated tetraalkyl ammonium salts, which contain more than 30 water molecules per molecule of the basic compound [33].

The first hypothesis seems improbable:

1. Deep imbedding of the ionic groups in the hydrophobic portion of the protein should prevent penetration of H^+ and OH^- ions.

Table 1. Maximum Possible Contribution of Hydrophobic Bonds
to Activation Energy of Denaturation ΔE_{max} and Experimental
Activation Energies (E_{eff}) at Neutral pH's for Certain Proteins

Protein	No. of hydrophobic groups	ΔE_{max}, kcal	E_{eff}, kcal	References
Human serum albumin	200	100	135	[42,43]
Egg albumin	120	60	180	[43,44]
Insulin.................	36	18	36	[22,43]
Pepsin.................	182	96	100	[43,45]
Chymotrypsinogen	76	38	80	[23,43]
Peroxidase, milk	320*	160	185	[22,46]
Catalyse, bovine liver	130	65	100	[43] †
Bacterial α-amylase	170	65	68	[43,47]
Peroxidase, horseradish	136	68	57	[46] ‡
Hemoglobin.............	232	116	76	[24,43]

*Calculated on the assumption that 40% of the functional groups are hydrophobic.

†Data obtained in conjunction with Yu. M. Azizov.

‡Data obtained by author.

2. Migration of ions from a nonpolar environment to water (Fig. 6) is accompanied by an enormous decrease in enthalpy, which materially exceeds the decrease in entropy [34]; migration of ions into water should therefore lead to high values of T_m (above 1000°K) and low values of E_{eff} and $\Delta S_{eff}^{\neq}$.

Conversely, formation of a bond by an ion in a hydrated fold provides access to neutralizing agents. As model experiments have shown, decomposition of ionic pairs in water can be accompanied by a decrease in enthalpy and entropy [35] (Fig. 6). There is a good approximation of a compensation relationship between ΔH and ΔS in this case, which apparently indicates that breakdown and formation of ionic pairs in water is accompanied by cooperative rearrangement of water molecules with $T_m \approx 300°K$.

There are also other arguments to support the hypothesis that ionic bonds play a major role, forming a single cooperative system with the hydrate–water molecules and perhaps with the adjacent hydrophobic groups. Numerous data on protein denaturation find a natural explanation within the framework of these concepts. The high values of E_{eff} and $\Delta S_{eff}^{\neq}$ at neutral pH's can be attributed to the need for "melting" of the surrounding water, which leads to high values of m and thus of E_{eff} and $\Delta S_{eff}^{\neq}$. The rapid neutralization of the ionic groups in acidic and alkaline media leads to conditions under which the cooperative water system can no longer be maintained by electrostatic forces. In kinetic terms, this means that $K_+ \gg K_-$ and $\overline{K} \ll 1$, and the process rate equals the initiation rate. In this case, E_{eff} and $\Delta S_{eff}^{\neq}$ should decrease to the levels of E_1 and $\Delta S_1^{\neq}$ [Eqs. (2) and (4)].

Calculation of the heat of solution of different amino acids in water–alcohol solutions* has shown that the effect of alcohol, which produces simultaneous increases in the ΔH and ΔS of solution, is similar to that of a rise in temperature. It can formally be assumed that both effects lead to "melting" of the hydrate water around the functional groups of the protein. Addition of alcohol substantially increases the E_{eff} and $\Delta S_{eff}^{\neq}$ of denaturation [22, 23], which can, by analogy with model systems, be attributed to the rise in ΔH_0 and ΔS_0 resulting from "melting" of the structured water. The ΔH and ΔS of ionic complexing in solution vary in direct proportion to the ionic strength [35] (Fig. 6). Addition of ions to certain proteins produces the same results [22, 23].

---

*The calculations were made with data on the solubility of amino acids in water and water–alcohol solutions at different values of T [36].

We should note still another interesting analogy. At certain temperatures between +10 and +30° (the so-called Bergman temperatures), aqueous solutions of a number of salts (KCl, KBr, KI, and $NH_4H_2PO_4$) undergo a phase transition that can be recorded from the change in heat capacity and the thermal coefficient of solubility [37-40]. It has been found that the influence of salts, organic solvents, urea, etc., on the Bergman temperature is similar to that of these compounds on the "melting" point of proteins.

It must be emphasized that there is no need for water to be stably bound into an ice-like structure for its cooperative properties to be manifested in the kinetic and thermodynamic properties of proteins. High E_{eff} and large thermal effects can be obtained at values of ΔH_0 close to or even less than kT, provided that the number of participants in the ensemble is sufficiently large. We can make the following rough estimate. The isotopic effect in the heats of fusion on moving from H_2O to D_2O is 80 cal [41]. Substitution of heavy for light water led to an isotopic effect in the E_{eff} of catalyse deactivation of 58,000 cal/mole [25]. If we assume that the principal contribution to the isotopic effect is made by the change in the heat of fusion of the hydrate water, we find that the state of about 700 water molecules should be altered during denaturation of one catalyse molecule. The contribution made to the kinetic parameters of denaturation by other types of bonds is evidently slight. Breakage of disulfide bonds requires an activation energy of 70-72 kcal, which depends little on the nature of the atom attached to the sulfur, and proceeds at a noticeable rate only at temperatures of 400-500°K [52]. The hydrogen bonds in water do not usually form extensive continuous cooperative regions and the energy effects of their breakage in water are small.

The conclusion that hydrated ionic groups make a material contribution to the stabilization of protein globules extends only to the few proteins considered in the present paper. We cannot discount the possibility that the principal contribution in other proteins is made by other bonds, including those of the hydrophobic type.

Experimental data on denaturation are usually described by Eyring's formula:

$$K = \frac{kT}{h} \exp\left(-\Delta H^{\neq}/RT + \Delta S^{\neq}/R\right), \tag{11}$$

where $\Delta H^{\neq}$ and $\Delta S^{\neq}$ are the activation enthalpy and entropy. However, the validity of direct application of the concepts of the absolute-rate theory to large cooperative ensembles is on the whole dubious. In particular, substitution of the experimental values of $\Delta H^{\neq}$ and $\Delta S^{\neq}$ into Eq. (11) in cases where these values are rather large produces a paradoxical result: the values of K may exceed 10^{13} sec^{-1} at comparatively low temperatures (T ≥ 350°K), i.e., the processes may take place at rates exceeding the rate of thermal movement. It was demonstrated above that the kinetic phenomena of denaturation are naturally explained by assuming that these processes have a multistage mechanism.

The author wishes to thank B. I. Sukhorukov for a fruitful discussion of the results of this work.

LITERATURE CITED

1. C. Nemethy, J. Steinberg, and H. Scheraga, Biopolymers, 1:43 (1963).
2. J. A. Sheellnian, Compt. Rend. Trav. Lat. Carlsberg, Ser. Chim., 29:223, 230 (1955).
3. B. H. Zimin and J. K. Brogg, J. Chem. Phys., 31:526 (1959).
4. M. V. Vol'kenshtein, Biofizika, 6:257 (1961).
5. J. Marmus, Dotu P., 183:1427 (1959).
6. R. M. Birshtein, Biofizika, 7:513 (1962).
7. O. B. Ptitsin and Yu. E. Eizner, Biofizika, 10:3 (1965).
8. S. Lifson and A. Roig, J. Chem. Phys., 34:1963 (1961).

9.   M. V. Vol'kenshtein, Yu. A. Gotlib, and O. B. Ptitsin, Fiz. Tverd. Tela (1961).

10. Yu. A. Gotlib, Fiz. Tverd. Tela, 3 : 2170 (1962).

11. M. V. Vol'kenshtein, I. M. Godzhaev, Yu. A. Gotlib, and O. B. Ptitsin, Uch. Zap. LGU,
     No. 4 (1961).

12. I. M. Godzhaev, Biofizika, 11 : 193 (1966).

13. B. I. Sukhorukov and G. I. Likhtenshtein, Biofizika, 10 : 935 (1965).

14. M. I. Temkin, Dokl. Akad. Nauk, SSSR, 152 : 615 (1965).

15. E. A. Pshenichnov, Dokl. Akad. Nauk SSSR, 166 : 1162 (1966).

16. M. I. Temkin, Dokl. Akad. Nauk SSSR, 165 : 615 (1965).

17. M. V. Vol'kenshtein and B. N. Gol'dshtein, Biochim. et Biophys. Acta, 115 : 471 (1966).

18. G. I. Likhtenshtein, Kinetika i Kataliz, 4 : 35 (1963).

19. G. I. Likhtenshtein, Dissertation [in Russian], Institute of Chemical Physics (1963).

20. Ya. S. Lebedev, Yu. A. Tsvetkov, and V. V. Voevodskii, Kinetika i Kataliz, 1 : 496 (1960).

21. G. I. Likhtenshtein and B. I. Sukhorukov, Zh. Fiz. Khim., 38 : 747 (1964).

22. H. Eyring and A. E. Stearm, Chem. Res., 24 : 253 (1939).

23. R. Lumru and H. Eyring, J. Phys. Chem., 58 : 110 (1954).

24. K. Okumiki, Advances in Enzymol., 23 : 29 (1961).

25. W. R. Guild and R. R. Tubergen, Science, 125 : 939 (1957).

26. R. B. Setlow and E. C. Pollard, Molecular Biophysics, Addison-Wesley, Reading, Mass. (1962).

27. S. E. Bresler and D. L. Talmud, Dokl. Akad. Nauk SSSR, 43 : 326, 367 (1944).

28. W. Kauzmann, Adv. Protein Chem., 14 : 1 (1959).

29. H. F. Fisher, Proc. Nat. Acad. Sci. USA, 51 : 1285 (1964).

30. H. A. Scheraga, in: The Proteins, Vol. 1 (H. Neurath, ed.), Academic Press, New York
     (1963), p. 478.

31. M. Laskowski and H. A. Scheraga, J. Am. Chem. Soc., 83 : 266 (1961).

32. P. L. Privalov, Biofizika, 8 : 308 (1963).

33. I. M. Klotz, in: Horizons in Biochemistry [Russian translation], Izd. Mir, Moscow
     (1964), p. 399.

34. B. E. Conway and J. O'M. Bockris, in: Some Problems in Contemporary Electro-
     chemistry [Russian translation], IL, Moscow (1958), p. 63.

35. F. Rossotti, in: Modern Chemistry of Coordination Compounds [Russian translation],
     IL, Moscow (1963), p. 13.

36. Handbook of Chemistry and Physics, 37th ed., Cleveland (1955), p. 1637.

37. A. G. Bergman and I. A. Vlasov, Dokl. Akad. Nauk SSSR, 34 : 64 (1942).

38. V. A. Polosin and M. I. Shakhporonov, Zh. Fiz. Khim., 13 : 53 (1939).

39. M. I. Shakhporonov, Dokl. Akad. Nauk SSSR, 78 : 323 (1951).

40. I. A. Vlasov and A. G. Bergman, Dokl. Akad. Nauk SSSR, 39 : 148 (1943).

41. C. Nemethy and H. A. Scheraga, J. Chem. Phys., 41 : 681 (1964).

42. R. J. Gibbs, Arch. Biochem. and Biophys., 51 : 277 (1954).

43. G. R. Tristam and R. H. Smith, in: The Proteins, Vol. 1 (H. Neurath, ed.), Academic
     Press, New York (1963), p. 46.

44. R. S. Lewis, Biochem. J., 20 : 978 (1926).

45. N. V. Kondakova, Dissertation [in Russian], Institute of Biophysics, Moscow (1964).

46. K. G. Paul, in: The Enzymes, Vol. 8 (P. D. Boyer, H. Lardy, and K. Myrbäck, eds.),
     Academic Press, New York (1963), p. 227.

47. E. H. Fischer and E. A. Stein, in: The Enzymes, Vol. 4 (P. D. Boyer, H. Lardy, and
     K. Myrbäck, eds.), Academic Press, New York (1960), p. 319.

48. D. French, in: The Enzymes, Vol. 4 (P. D. Boyer, H. Lardy, and K. Myrbäck, eds.),
     Academic Press, New York, (1960), p. 349.

49. H. Lineuweaker and S. Schwimmer, The Enzymes, Vol. 10 (1941), p. 81.

50. A. Makagama and H. A. Scheraga, J. Am. Chem. Soc., 83 : 1575 (1961).

51. A. Loeb and H. A. Scheraga, J. Am. Chem. Soc., 84 : 134 (1962).

INFLUENCE OF SHORT-RANGE FORCES ON THE THERMODYNAMICS OF STRONG ELECTROLYTE SOLUTIONS AND THEIR RELATIONSHIP TO SOLUTION STRUCTURE

Yu. M. Kessler and R. I. Mostkova

Institute of Electrochemistry
Academy of Sciences of the USSR

The complexity of systematic statistical calculation of the thermodynamic properties of strong electrolyte solutions, taking proper account of the solvent, forces us to resort to certain simplifications. In the traditional approach, the solvent is assigned the role of a dielectric continuum. This reduces the purely computational difficulties, which become almost insurmountable, but raises the question of the existence of effects not covered by the approximation and of techniques for investigating them. There are accordingly two current problems in the theory of moderately concentrated strong electrolyte solutions: 1) correct theoretical calculation of the electrostatic (coulombic) free energy for the classical model of "rigid charged spheres in a continuous medium"; 2) determination of the nature and contribution of noncoulombic effects and of the manner in which they should be evaluated. The present paper is devoted to certain aspects of the second problem.

The solutions to these problems are interrelated. It would be rather simple to determine the physical nature of the noncoulombic contributions and make a quantitative evaluation if there were a precise solution to the classical model; it is almost impossible to do so in the absence of such a solution. The situation actually lies between these extremes, since all existing solutions for final concentrations must be regarded as approximate. However, they all describe experimental results equally well through the use of the adjusted parameter r_0, which is a component of all the theories and has the sense of the distance of closest ion approach. Each theory naturally requires a different value of r_0 for a given subject. At the same time, the variation in r_0 for a given electrolyte in different solvents and the difference

$$\Delta r = r_0 - r_{cr} \tag{1}$$

$$\Delta r_1 = (r_0) \ \text{solvent}_1 - (r_0) \ \text{solvent}_2 \tag{2}$$

(where r_{cr} is the sum of the crystallographic radii of the anion and cation) can be used, at least in principle, to evaluate the character of the changes in and differences between the true interionic potential and the coulombic potential. Analysis of the existing theoretical equations showed that almost all of them yield a set of r_0 suitable for this purpose, with the exception of the second-approximation equation in the Debye−Hueckel theory, in deriving which the authors omitted a term to provide a correction for the true ionic volume. The Mayer−Haag and

Levich−Kir'yanov equations are the most rigorous. The former yields an r_0 about twice as large as the latter. We employed both equations in calculating r_0, in order to make a more objective judgment.

The difference between the behavior of real solutions and that theoretically predicted for the classical model results from the difference between the interionic interaction potential $U_{ab}(r)$ and the coulombic potential $U_{cou}(r)$ at short interionic distances. Since the theories discussed above contain only one adjusted parameter (r_0), there should be a correlation such that

$$U_{ab}(r) > U_{cov}(r) \quad \text{at} \quad \Delta r > 0$$

$$[U_{ab}(r)]_{svt_1} > [U_{ab}(r)]_{svt_2} \quad \text{at} \quad (r_0)_{svt_1} > (r_0)_{svt_2}$$

$$U_{ab}(r) < U_{cov}(r) \quad \text{at} \quad \Delta r < 0$$

$$[U_{ab}(r)]_{svt_1} < [U_{ab}(r)]_{svt_2} \quad \text{at} \quad (r_0)_{svt_1} < (r_0)_{svt_2}.$$

(3)

The tendency toward association (in the sense of formation of short-lived ionic pairs) usually decreases as Δr and Δr_1 increase.

Let us consider the factors governing $U_{ab}(r)$, the manner in which it is to be evaluated, and the solution properties with which it is correlated and touch on certain physical aspects of the concept of association. We introduce the parameter r_s, which is defined as the sum of the radii of the spheres of primary ionic solvation. The Levich−Kir'yanov equation defines the free energy of an electrolyte solution as the sum of the coulombic component produced by ionic interaction at $r \geq r_s$ and the term produced by interaction at small $r_{cr} \leq r < r_s$:

$$\Delta F = \Delta F_{cou} - c^2 \sum_{a,b} B \int_{r_{cr}}^{r_s} \exp\left[-G_{ab}(r)\right] \left[\exp\left(-\frac{U_{a,b}(r)}{kT} - 1\right) \right] r^2 \, dr = \Delta F_{cou} - \Delta F_s$$

(4)

$$G_{ab}(r) = \frac{z_a z_b e^2}{DkTr} f(r_s, r, c); \quad f(r_s, r, c) \simeq 1 - 0.8$$

(5)

with c = 0.01−0.10 mole/liter. The quantity $U_{ab}(r)$ can be represented as the sum of the coulombic and noncoulombic constituents:

$$U_{ab}(r) = \frac{z_a z_b e^2}{Dr} + U_{ab}^{spec} |r| = \frac{-|z_a z_b| e^2}{Dr} + \Delta U_{ab}^s(r) + f_{ab}(r),$$

(6)

where $\Delta U_{ab}^s(r)$ is the difference between the solvation energies of the ionic pair and the free ions; $f_{ab}(r)$ takes into account all other effects, such as the change in the electrostatic energy of ionic interaction resulting from dielectric saturation and the quantum-mechanical interaction at very small r.

A change in the value of ΔF_s is a macroscopic manifestation of a change in the specific interionic interaction. The literature usually relates this change to the variation in $U_{ab}(r)$. However, this interpretation is not accurate if we are dealing with different solvents having different D, since it can be seen from Eq. (6) that $U_{ab}(r)$ can remain invariant as a result of random compensation of its components, while $G_{ab}(r)$ is altered simply by virtue of the change in the product DT, which also causes a change in ΔF_s in accordance with Eq. (4). In the general case, it is therefore impossible to conclude that $U_{ab}(r)$ changes merely because ΔF_s varies.

Furthermore, it is now widely believed that ionic association decreases as the product DT and the free-ion solvation increase. This is also invalid in the general case, provided that

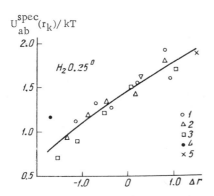

Fig. 1. Average specific energies of interionic interaction $U_{ab}^{spec}[r_k]/kT$ as a function of Δr in water. 1) Chlorides; 2) bromides; 3) iodides; 4) nitrates; 5) alkali-metal hydroxides, 25°C.

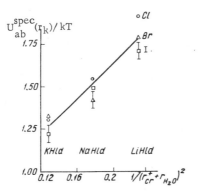

Fig. 2. $U_{ab}^{spec}[r_k]/kT$ as a function of energy of ion−dipole interaction of cations for lithium, sodium, and potassium halides in water, 25°C.

the physical premises on which Eq. (6) is based are true. The second and third terms in Eq. (6) actually depend very little on D. At the same time, ion-pair solvation should depend to a large extent on the structural characteristics of the solvent, particularly the lability of the liquid structure. The more labile this structure, the less energy is expended in reorientation of the solvent molecules solvating the ionic pair. The decrease in $\Delta U_{ab}^{S}(r)$ can theoretically exceed the increase in the first term of Eq. (6) produced by the rise in D; an increase in D and free-ion solvation can therefore be accompanied by an increase in the number of ionic pairs. The literature has paid very little attention to the role of ion-pair solvation.

In seeking experimental confirmation of the foregoing, we employed the data in the literature [1] and measured the activity coefficients of sodium, potassium, and cesium chlorides at 25°C in mixtures of formamide and 40% by weight acetamide (D = 95), formamide (D = 109), and N-methyl formamide (D = 182), and at 2.5°C in formamide (D = 118), since the only correct data in the literature for solvents with high D are for three salts in prussic acid at −13°C (D = 206) and for HCl in N-methylacetamide at 40°C (D = 165). We also determined the change in the standard free energy of solvation $\Delta F_S = \Delta F_{H_2O} - \Delta F_{svt}$, as well as the molar volumes and equivalent electrical conductivities at two concentrations for a mixture of formamide and acetamide. We calculated r_S from Eq. (4) and then found the effective values of $U_{ab}|r|$ and $U_{ab}^{spec}|r|$ from the formula:

$$\int_l^m f|x|\varphi|x|\,dx = f|x_k|\int_l^m \varphi(x)\,dx, \qquad (7)$$

and calculated r_0 from Eq. (4), setting $\Delta F_S \equiv 0$ (i.e., $r_S = r_{cr}$). Our first calculations were made for water at 25°C with 20 electrolytes and at 0°C with 8 electrolytes; we checked for a correlation between $U_{ab}^{spec}(r)$ and Δr and confirmed its existence (Fig. 1). Since it is substantially more cumbersome to calculate Δr than $U_{ab}^{spec}(r_k)$, this result is of practical interest, permitting rapid qualitative evaluation of the change in $U_{ab}^{spec}(r)$ from the change in Δr.

The relationship between the change in $U_{ab}^{spec}(r)$ and the properties of the ions is of particular interest. Functions showing the influence of the difference in the solvation energies of the ionic pair and of the free ions $\Delta U_{ab}^{S}(r)$ on the probability of direct contact were found for aqueous solutions (Figs. 1 and 2). The curves in Figs. 2 and 3 can be generalized by constructing a function representing the change in ion-pair energy as the interionic distance is varied from $r = r_S$ to $r = r_{cr}$, having the form

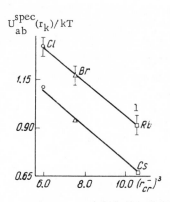

Fig. 3. $U_{ab}^{spec}[r_k]/kT$ as a function of anion polarizability for rubidium and cesium halides in water, 25°C.

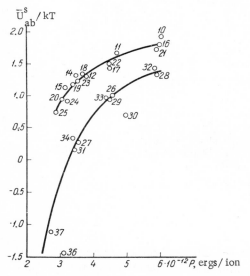

Fig. 4. $U_{ab}^{spec}[r_k]/kT$ as a function of change in ion energy during approach for I−I electrolytes in water, 25°C.

$$P_1 = -\frac{(z_1 z_2)e^2}{D_{eff}}$$

$$\int_{r_{cr}}^{r_s} \frac{dr}{r^2} + \mu e \left(\frac{n_+}{(r_{cr}^+ + r_p)^2} + \frac{n_-}{(r_{cr}^- + r_p)^2} \right) - U_{g\cdot g}, \quad (8)$$

where the value

$$D_{eff} = 5 \exp(r_{cr}/2r_{H_2O}), \quad (9)$$

is assumed for the effective value of D near the ion, μ is the dipole moment of water, and n_+ and n_- are the numbers of water molecules displaced from the ionic solvation spheres by the ionic overlap (calculated from the corresponding overlap-region volume ΔV_i, the coordination number of the ion in question n_k^i, and the volume of the solvate sphere one solvent molecule thick around the ion V_s^i)

$$n_i = \frac{n_k^i}{V_s^i} \Delta V_i, \quad (10)$$

and $U_{g\cdot g}$ is the energy of the dispersion interaction of ions in direct contact. It can be seen from Fig. 4 that the short-range interaction is governed to a substantial extent by these effects. These results are in agreement with the fact that r_0 increases from salt to salt over the series LiCl− LiBr− LiI, NaCl−NaBr−NaI, and KCl−KBr−KI by an amount approximately equal to the difference in the crystallographic radii of the anions, while this phenomenon is not observed for Rb and Cs salts (Table 1). The general regularities observed for aqueous solutions thus result from the differing degrees of hydration, principally cationic.

Let us now consider the change in the properties of the same electrolyte in different solvents. The solvents in Table 2 are arranged in order of increasing values of the product DT. It can be seen from this table that the tendency toward formation of ion pairs is greater in the mixtures than in water, although the solvation is also greater in the former case. Moreover, ion-pair formation in formamide is substantially more pronounced at 25°C than at 2.5°C, although the free−ion solvation changes very slightly and DT remains almost constant. In both cases, the only possible cause of the increase in the probability that the ions will remain at short distances is the corresponding change in ion−pair solvation energy. On the other hand, the increase in this probability for HCN results from a decrease in free−ion solvation, since HCN is known to poorly solvate alkali-metal ions. Calculation of the effective potentials yields a more detailed picture; in the cases investigated (Table 3), $U_{ab}^{spec}(r_k)$ decreases as D and the free−ion solvation increase, which also points to the shift in ion−pair solvation energy toward more negative values as the cause of the decrease. As was pointed out above, this shift may be due to an

Table 1. Hydration Parameters of Certain Cations

H_2O	25°	M	Li	Na	K	Rb	Cs
$r_{cr}^{I^-} - r_{cr}^{Cl^-}$	0.39	$(r_0)_{MI} - (r_0)_{MCl}$	0.60	0.25	0.21	—0.06	—0.07
$r_{cr}^{Br^-} - r_{cr}^{Cl^-}$	0.15	$(r_0)_{MBr} - (r_0)_{MCl}$	0.15	0.11	0.05	—0.03	—0.07

Table 2. Change in Electrolyte Properties in Different Solvents, r_0

Solvent	H_2O, 25°	H_2O, 0°	Formamide + acetamide, 25°C	Formamide, 25°C	Formamide, 2.5°C	HCN, —13.3°	Methylformamide, 25°C
DT · 10^{-4}	2.34	2.40	2.85	3.27	3.27	5.35	5.44
NaCl	3.02	3.16	2.44	3.03	4.64	—	3.9
KCl	2.81	2.72	—	—	4.60	—	3.8
CsCl	2.35	—	1.90	2.56	3.40	—	3.5
KI	3.02	—	—	—	4.14	1.70	—

Table 3. Effective Potentials

Electrolyte	$\dfrac{U_{ab}(r_K)}{\kappa T}$	$\dfrac{U_{ab}^{spec}(r_K)}{\kappa T}$	$\dfrac{U_{ab}(r_K)}{\kappa T}$	$\dfrac{U_{ab}^{spec}(r_K)}{\kappa T}$	$\dfrac{U_{ab}(r_K)}{\kappa T}$	$\dfrac{U_{ab}^{spec}(r_K)}{\kappa T}$	$\dfrac{U_{ab}(r_K)}{\kappa T}$	$\dfrac{U_{ab}^{spec}(r_K)}{\kappa T}$
	H_2O		Formamide + acetamide		Formamide		Methylformamide	
NaCl	—0.33	1.63	—0.87	1.19	—0.29	0.85	—0.02	0.43
CsCl	—0.70	1.11	—0.86	0.63	—1.11	0.27	—0.07	0.24

Note. The value of DT · 10^{-4} is 2.34 for H_2O, 2.85 for formamide + acetamide, 3.27 for formamide, and 5.44 for methylformamide.

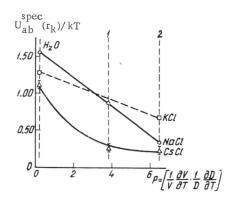

Fig. 5. $U_{ab}^{spec}[r_k]/kT$ as a function of p in water, formamide (1), and methylacetamide (2) for sodium, potassium, and cesium chlorides, 25°C.

increase in the lability of the solvent structure. The latter can be characterized by the product of the thermal coefficients of volume and D:

$$p = \frac{1}{VD}\frac{\partial V}{\partial T}\frac{\partial D}{\partial T}. \qquad (11)$$

Actually, the graph representing $U_{ab}^{spec}(r_k) = f(p)$ shows that there is a correlation between these quantities (Fig. 5).

It must be noted that $\Delta U_{ab}^{S}(r)$ varies greatly in accordance with the direction in which the solvent molecules are displaced when two ions approach. The ratio of the radii of the solvent anion and cation are also important in this case. If they are equal, the ions come into contact as they approach and the two solvent molecules displaced (one from the cation and one from the anion) simultaneously touch both ions (symmetrically with respect to the line joining their centers); the desolvation energy is then 0.4U, where U is the total energy of the ion−dipole interaction of the cation and anion with a solvent molecule, i.e., the solvation energy is more favorable for the ion pair than for the free ions. If $r_{cr} = \infty$, $\Delta U_{ab}^{S}(r) = 0.5$ U; finally, if one of the two displaced solvent molecules enters the second coordination sphere as the ions approach, the desolvation energy is about +0.2U (if the cationic and anionic radii are equal). It follows from the conductivity data in the literature that the association of potassium perchlorate is markedly greater in aqueous solutions than in dimethylformamide, dimethylacetamide, dimethylpropionamide, or dimethylsulfoxide, for which $D \cong 0.5D_{H_2O}$. Moreover, the association of LiCl and LiNO$_3$ is greater in HCN (18°C, D = 118) than in H$_2$O at the same temperature. The reason for the latter phenomenon is apparently a weakening of the solvation of the Li$^+$ ion, whose mobility in HCN is greater than that of Na$^+$, while the situation is reversed in water.

The foregoing material thus demonstrates the need to give equal consideration to the solvation energies of both free ions and ion pairs and confirms our previous hypotheses regarding the physical nature of the interionic interaction potential at small r.

LITERATURE CITED

1.   R. A. Robinson and R. Stokes, Electrolyte Solutions, Butterworths Scientific Publishers, London (1959).

PARTICIPATION OF H_2O (OH^- IONS) IN OXIDATION—REDUCTION PROCESSES

G. V. Fomin, L. A. Blyumenfel'd, R. M. Davydov, and L. G. Ignat'eva

Institute of Chemical Physics
Academy of Sciences of the USSR

It is well known that water takes an active part in electron-transfer processes during photosynthesis, since the OH^- ion is the immediate donor of the electron that ultimately enters the electron vacancy in the oxidative center produced after excitation of chlorophyll. At the same time, when dark biochemical processes in plant and animal cells are considered, water is usually assigned the role of a solvent, i.e., of a structuring factor or, at best, of the medium in which charge or excitation migration takes place, as in the theories of Klotz and Szent-Györgyi [1, 2]. The present paper summarizes the results of investigations conducted in our laboratory over the past few years with certain model systems, on the basis of which we can pose the question of whether water, or, more precisely, the OH^- ion, cannot participate directly in oxidation—reduction processes and in dark reactions.

Let us examine some experimental results. We reported in a previous article [3] that during solution of a strong electron acceptor, such as tetracyanethylene (TCE), in water or a water—alcohol mixture at neutral pH's, free radicals develop in the solution. On the basis of their EPR spectra, these can be identified as TCE ion-radicals. The radical yield increases with the solution pH and the latter decreases in such fashion that the number of hydroxyl ions that disappear is approximately equal to the number of TCE molecules added. Similar results were obtained for a whole series of electron acceptors, such as chloranil, n-benzoquinone, naphthoquinone, duroquinone, methylene blue, tetracyanquinodimethane, etc. The majority of acceptors form radicals at substantially alkaline pH's. This apparently results not so much from the increase in radical-formation rate at high hydroxyl-ion concentrations as from the high radical-decay rate, since the semiquinone radicals of most acceptors exist in protonized form at neutral pH's and the rate constant of the disproportionation reaction is several orders of magnitude higher than the corresponding constant for the ion-radicals. The fact that this reaction occurs with chemical compounds of differing structure having only one property in common (a relatively high electron affinity), as well as certain general considerations, forced us to assume the following reaction pattern:

$$OH^- + A \rightarrow OH + A^-$$
$$4\,OH \rightarrow 2\,H_2O + O_2, \tag{1}$$

where A is an electron acceptor.

System I is not infeasible on the basis of general considerations. For example, the redox potentials for the OH$^-$/O$_2$ and I$^-$/I$_2$ systems are given as +0.401 and +0.536, respectively [4]. Moreover, it has been proved [5] that the reaction of I$^-$ with TCE, chloranil, or n-benzoquinone has the form

$$I^- + A \rightarrow I + A^-$$
$$2\,I \rightarrow I_2$$

Furthermore, if we use the known ionization potential of the OH$^-$ ion (1.73 eV [6]) and the electron affinity of TCE (2.8 eV [7]) in the gaseous phase, reaction (I) should be thermodynamically advantageous. It is difficult to take into account the influence of the solvent, but even if the reaction in solution is not thermodynamically advantageous, formation of extremely active OH radicals will displace the equilibrium point and ensure that the process takes place.

The literature [8–10] proposes another pattern for the reaction of a hydroxyl ion with n-benzoquinone or methylnaphthoquinone (quinones that contain labile hydrogen):

(II)

One proof of this pattern is the fact that the corresponding hydroxyquinones are found in the reaction mixture. In our opinion, however, hydroxyquinones can also be formed during the reaction of OH radicals with quinones or semiquinones. In addition, it has been demonstrated [11] that hydroxyquinones are formed during the reaction of quinones with hydrogen peroxide (which can appear in the system as a result of secondary reactions):

Reaction II also requires the presence of labile hydrogen for radicals to be formed. We have experimentally observed radical formation in acceptors not containing labile hydrogen atoms.

Detection of OH radicals during the course of the reaction might aid us in making a definite choice between systems I and II. However, their extremely high chemical activity makes this scarcely possible at present. One might expect to find hydrogen peroxide in the reaction mixture, but, first of all, it is hard to anticipate any large H$_2$O$_2$ yield since, according to Kazarnovskii [14], OH radicals react principally by disproportionation to form water and atomic oxygen and, in addition, they disappear in their reactions with quinones and semiquinones; secondly, as we previously demonstrated [13] and as will be discussed in this article, hydrogen peroxide is consumed during its interaction with semiquinones. The fact that we previously detected traces of hydrogen peroxide [3] is not a reliable proof of system I, since the amounts of

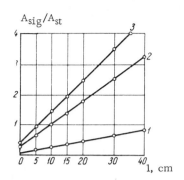

Fig. 1. Kinetic curves for formation of semiquinone radicals in solution of quinone and alkali saturated with: 1) oxygen; 2) air; 3) argon; $[OH^-] = 10^{-1}$ mole/liter; $[X] = 1.25 \cdot 10^{-3}$ mole/liter.

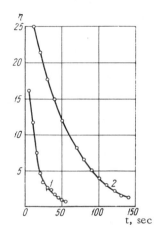

Fig. 2. Kinetic curves for decay of semiquinone radicals in solutions saturated with O_2 (1) and Ar (2). $[OH^-] = 10^{-1}$ mole/liter; $[X] = 1.25 \cdot 10^{-3}$ mole/liter.

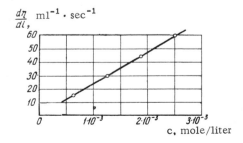

Fig. 3. Reaction rate as a function of quinone concentration. $[OH^-] = 10^{-1}$ mole/liter.

H_2O_2 involved could have been formed as a result of oxidation of the semiquinones by traces of oxygen in the reaction mixture. Moreover, it is also difficult to detect oxygen evolution, since (as will be shown below) oxygen reacts with the semiquinone radicals formed during the reaction. It is known that n-benzoquinone is oxidized to oxalic acid and CO_2 by molecular oxygen in alkaline media [12].

It is thus difficult to draw any conclusion regarding the actual reaction mechanism from the available data. We became interested in conducting a series of kinetic experiments that confirm or disprove one reaction pattern or the other. It must be noted that the literature contains no reports of detailed kinetic studies of these reactions. Reactions involving most of the acceptors investigated proceed extremely rapidly, so that only secondary radical-decay reactions can be observed by the usual technique of recording the EPR spectra of the reacting mixture; our main experiments were therefore conducted by the jet method, which permitted us to trace the kinetics of radical formation starting at a point 30 msec after the reaction began. The acceptor and alkali solutions were mixed in front of the resonator and the reacting mixture flowed through a capillary passing through the resonator. By observing the change in radical concentration along the capillary, we were able to follow the reaction kinetics. The kinetic curves, which will be given later, were plotted as a function of the distance from the reagent-mixing point to the resonator cover. The origin of the abscissa corresponds to 30 msec after the reaction began and each 5 cm on the time scale represents 50 msec.

Our main kinetic measurements by the jet method were made with n-benzoquinone (X) in aqueous solutions (which enabled us to exclude the effects associated with use of mixed solvents). The experiments were carried out at $[OH^-] \gg [X]$; the kinetic radical-formation curve was described by the monomolecular-reaction equation in this case. The solutions were saturated with argon in all the measurements. Oxygen in the solution affects both the radical-formation and radical-decay rates. Figures 1 and 2 show the corresponding experimental curves. The kinetics of radical decay were traced by the arrested-jet method. The

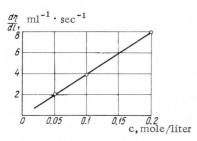

Fig. 4. Reaction rate as a function of hydroxyl-ion concentration. [X] = $1.25 \cdot 10^{-3}$ mole/liter.

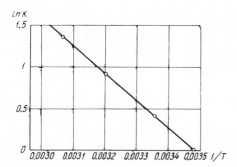

Fig. 5. Reaction rate as a function of temperature (on Arrhenian coordinates). $[OH^-] = 10^{-1}$ mole/liter; [X] = $2.5 \cdot 10^{-3}$ mole/liter.

increase in radical-formation rate in solutions saturated with argon or air requires special investigation: it may be due to the known structuring effect of argon and nitrogen (air) on water. The increase in the decay rate is evidently caused by the oxidation reactions discussed by Kawasaki [12].

By making kinetic measurements at different component concentrations, one can determine the order of the corresponding reaction with respect to the components. It can be seen from Figs. 3 and 4 that the reaction under investigation is first-order with respect to both quinone concentration and hydroxyl-ion concentration.

The temperature function of the reaction rate is well described by the Arrhenian equation (Fig. 5). The calculated activation energy E_a is 8.5 kcal/mole and the calculated pre-exponential factor K_0 is about 10^7 liters/mole · sec.

Kinetic measurements can also be made potentiometrically, tracing the change in hydroxyl-ion concentration. We made such measurements at pH 9 and a quinone concentration [X] = 10^{-2} mole/liter, so that $(OH^-) \ll (X)$. This technique was also used to evaluate the activation energy of the quinone−alkali reaction, but such experiments are of lesser accuracy; the value obtained for E_a was 8-10 kcal/mole.

Temperature−EPR measurements were also made with duroquinone. This compound is a substantially weaker acceptor than quinone and the reaction accordingly proceeds more slowly; we were able to make measurements without using the jet method and found the activation energy $-E_a$ to be 20 kcal/mole.

According to system I, hydrogen peroxide can be formed during the reaction of quinone with an alkali by the successive reactions O + OH → HO_2 and $2HO_2$ → H_2O_2 + O_2 [15]. We became interested in studying the reaction of H_2O_2 with semiquinone radicals and determining whether peroxide accumulation can be expected in such reactions. We had previously studied the reaction of hydrogen peroxide with the semiquinone ion-radicals of ditributyl quinone and demonstrated that ion-radical decay in this reaction is roughly first-order with respect to H_2O_2 [13]. The reaction was carried out at $(OH^-) = 10^{-2}$ mole/liter in a water−alcohol medium (1:10). In the present work, we investigated the influence of hydrogen peroxide added to the system on the kinetics of both formation and decay of n-benzoquinone semiquinones. It was found that the effect of H_2O_2 on the reaction kinetics is governed both by its concentration and by the solution pH. We studied the influence of different H_2O_2 concentrations on the course of the reaction at different medium pH's. Figure 6 shows kinetic curves for semiquinone−radical formation at different H_2O_2 concentrations and $(OH^-) = 5 \cdot 10^{-2}$ mole/liter. The initial radical-formation rate increased with the hydrogen peroxide concentration. The inflections in the curves evidently result from consumption of the peroxide. Considering that the dissociation constant K of hydrogen peroxide ($H_2O_2 \rightleftharpoons HO_2^- + H^+$) is $1.4 \cdot 10^{-12}$ [15], we can assume that reactions associated with the reducing properties of HO_2^- ions are responsible for the increase in radical-formation rate.

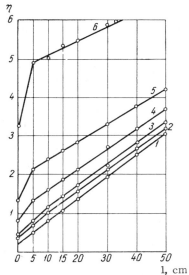

Fig. 6. Kinetic curves for radical formation at different hydrogen peroxide concentrations in solution of quinone and alkali. $(OH^-) = 10^{-1}$ mole/liter; $(X) = 1.25 \cdot 10^{-3}$ mole/liter (pH > $pK_{H_2O_2}$). 1) $(H_2O_2) = 0$; 2) $(H_2O_2) = 5.6 \cdot 10^{-6}$ mole per liter; 3) $(H_2O_2) = 1.4 \cdot 10^{-5}$ mole/liter; 4) $(H_2O_2) = 2.8 \cdot 10^{-5}$ mole/liter; 5) $(H_2O_2) = 5.6 \cdot 10^{-5}$ mole/liter; 6) $(H_2O_2) = 1.7 : 10^{-4}$ mole/liter.

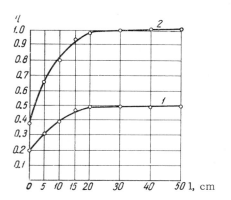

Fig. 7. Kinetic curves for radical formation in presence of H_2O_2 at pH < $pK_{H_2O_2}$. $(OH^-) = 10^{-3}$ mole/liter; $(X) = 5 \cdot 10^{-3}$ mole/liter. 1) $(H_2O_2) = 1.7 \cdot 10^{-2}$ mole per liter; 2) $(H_2O_2) = 0$.

When the reaction was conducted at a pH below the pK of hydrogen peroxide, the H_2O_2 added caused a decrease in the reaction rate (Fig. 7). The reaction rate was reduced in alcoholic solution and at $(OH^-) = 5 \cdot 10^{-2}$ mole per liter, apparently as a result of the decrease in the dissociation constant of H_2O_2 in alcoholic solutions. Figure 8 shows kinetic curves for the decay of semiquinone radicals at different hydrogen peroxide concentrations. The reaction was conducted at a pH below the pK of H_2O_2. There was a marked increase in the radical-decay rate at $(H_2O_2) > (X)$.

Hydrogen peroxide is thus consumed in both radical-decay and radical-formation reactions; as has already been noted, it is therefore difficult to anticipate any pronounced H_2O_2 accumulation in such reactions and detection of H_2O_2 cannot be used as proof of system I. The results of our kinetic measurements (the simple shape of the kinetic curve, the fact that the reactions were first-order with respect to components, and the parallelism of the reaction rate and the electron affinity of the acceptor) are in conformity with system I. However, these experiments still do not provide us with grounds for conclusive choice of a reaction pattern, since the same kinetic regularities can be expected when the constants in system II have a definite ratio.

It would be helpful to show that the kinetic regularities observed for n-benzoquinone are common to many acceptors, but the decisive experiment would naturally be detection of OH radicals in such reactions.

The active centers of many enzymes in biological systems are electron acceptors of the quinone or flavin type. A similar interaction between these acceptors and hydroxyl ions can presumably occur in biochemical processes. Actually, certain experiments conducted by the steam-jet method [16] force us to assume that at least some of the EPR signals observed in metabolizing tissues are produced by reactions in which hydroxyl ions participate [13].

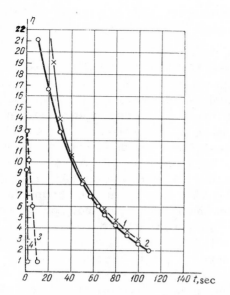

Fig. 8. Kinetic curves for decay of semi-quinone radicals in presence of H_2O_2. $(OH^-) = 10^{-1}$ mole/liter; $(X) = 1.25 \cdot 10^{-3}$ mole/liter. 1) $(H_2O_2) = 0$; 2) $(H_2O_2) = 5.6 \cdot 10^{-4}$ mole/liter; 3) $(H_2O_2) = 5.6 \cdot 10^{-3}$ mole/liter; 4) $(H_2O_2) = 8.4 \cdot 10^{-2}$ mole per liter.

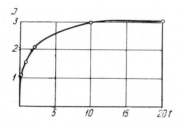

Fig. 9. Kinetic curve for radical formation in solution of hemoglobin and quinone (in buffer at pH 7); the time is in minutes. $(X) = 4 \cdot 10^{-3}$ mole/liter; (hemoglobin) = $5 \cdot 10^{-5}$ mole/liter.

At the same time, no measurable amounts of radicals are formed in solutions of such acceptors as quinones and flavins at physiological pH's. We previously suggested [13] that the shift in the reaction toward lower pH's might be caused by an increase in the radical-formation rate at the biopolymer surface as a result of the anomalously high hydroxyl-ion mobility in the layer of structured water at the surface (Grotgus mechanism) and from stabilization of the radicals at the surface.

We became interested in studying the reaction of electron acceptors with water (OH^-) in the presence of proteins. It was found that radicals are produced at physiological pH's in this case, while no detectable amounts of radicals are formed on a neutral base (quartz powder). The EPR signal of radicals sorbed on proteins is a singlet 6-8 Oe wide, similar to that observed by Kharitonenkov [17] during oxidation of various hydroquinones on a protein base. Our experiments were conducted in the following manner. Solutions of a protein (in a buffer at pH 7.5-8) and an acceptor (n-benzoquinone or β-naphthoquinone) were mixed, incubated for a predetermined period, frozen, and lyophilically dried. Figure 9 shows the kinetics of accumulation of n-benzoquinone radicals in the presence of hemoglobin. The signal intensity increased for about 10 min and then remained almost constant for about 1 h.

We investigated the radical yield, using different proteins as bases. Table 1 presents the data obtained. This table also shows certain of the characteristics of the proteins investigated. The radical concentrations were calculated per gram of protein. It must be noted that, as for the hydroquinone—semiquinone conversion in the presence of proteins [17],

Table 1. Radical Yield and Characteristics of Proteins

Protein	$I \cdot 10^{-16}$	Isoelec. point	Proportion of α-helix, %	No. of groups
Hemoglobin	20	6.7	75—80	2
Serum albumin	8	5.0	46—58	17
Trypsin	5	10.1	—	—
Urease	5	—	—	—
Globin	5	7.5	48	—
Egg albumin	3	4.6	31—50	5
Pepsin	1.3	1.0	26—31	0

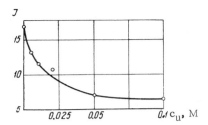

Fig. 10. Radical yield in solution of hemoglobin and β-naphthoquinone at different urea concentrations; (β-naphthoquinone) = $4 \cdot 10^{-4}$ mole/liter; (hemoglobin) = $5 \cdot 10^{-5}$ mole/liter.

we observed a definite correlation between the radical yield and the proportion of α-helix in the protein.

In addition to the proposed mechanism of radical formation by reaction of the acceptor with hydroxyl ions, we must consider the possibility of a direct reaction between the protein and the acceptor. As was previously demonstrated [18], we must take into account principally the interaction between the acceptor and the cysteine residue of the protein. However, as can be seen from our experiments on the blocking of SH groups by n-chloromercuribenzoate [18] and from the lack of any correlation between the SH content of the protein and the radical yield, direct reaction between acceptor and protein can account for only a small portion of the semiquinone radicals that develop in the presence of the latter. The slight decrease in radical yield when the SH groups are blocked is apparently due not to electron transfer from the protein (SH groups) to the quinone but to a change in protein conformation, since a similar drop in radical yield occurs during oxidation of hydroquinones on a protein with blocked SH groups [18]. If we assume that one of the factors responsible for shifting the radical-formation reaction toward neutral pH's in the presence of protein is the increase in radical-formation rate resulting from the increase in hydroxyl-ion mobility in the layer of structured water around the biopolymer, such agents as urea and its derivatives (guanidine), which disrupt the water structure, should cause a decrease in radical yield. Figure 10 presents the results of pertinent experiments. There was actually a noticeable drop in radical yield at low urea concentrations. Similar results were obtained for guanidine. It must be noted that the urea and guanidine concentrations studied have no denaturative effect on proteins [19] (moreover, denaturation does not markedly affect the reaction in question [18]). The kinetics of radical decay in alkaline aqueous solutions also remained unchanged in the presence of guanidine and urea (these experiments were conducted by the jet method). It is difficult to imagine that the effect of these compounds on the radical-decay reaction would be different in the presence of protein or at neutral pH's. In our opinion, the most probable explanation of such effects lies in the action of such compounds on the structure of water.

Conclusions

Our experimental results force us to draw the following main conclusions.

The hydroxyl ion can participate in oxidation–reduction reactions with organic electron acceptors, ultimately being an electron donor. The detailed mechanism of this process (direct transfer from OH$^-$ to the acceptor or preliminary attachment of OH$^-$ to the acceptor) requires clarification. However, a number of preliminary data impel us to adopt the first alternative as a preliminary hypothesis. Redox processes involving typical biochemically important organic electron acceptors are displaced toward physiological pH's in the presence of proteins. A number of data indicate that this is caused by an increase in the process cross section resulting from charge migration from OH$^-$ to the electron acceptor through the layer of structured water at the biopolymer surface, by the Grotgus mechanism. These results can be regarded as justification for the aforementioned hypothesis that water (OH$^-$ ions) participates directly in formation of electron–acceptor ion-radicals and, in general, in the redox processes involved in intracellular dark reactions. It can be surmised that the mechanism by which water participates in dark biochemical processes has much in common with that of the photooxidation of water during photosynthesis.

LITERATURE CITED

1. I.M. Klotz, in: Horizons in Biochemistry [Russian translation], Izd. Mir, Moscow (1964).
2. A. Szent-Györgyi, Bioenergetics, Academic Press, New York (1957).
3. G. V. Fomin, L. A. Blyumenfel'd, and B. I. Sukhorukov, Dokl. Akad. Nauk SSSR, 157(5): 1199 (1964).
4. L. I. Antropov, Theoretical Electrochemistry [in Russian], Izd. Vyssh. Shkola, Moscow (1965).
5. O. W. Webster, W. Manler, and R. E. Benson, J. Org. Chem., 25:1 470 (1960).
6. V. I. Vedeneev, L. V. Gurvich, et al., Rupture Energy of Chemical Bonds: Ionization Potentials and Electron Affinity [in Russian], Izd. AN SSSR, Moscow (1962).
7. M. Battley and L. E. Lyens, Nature, 196(4854):753 (1962).
8. M. Eigen and P. Matthies, Chem. Ber., 94:3309 (1960).
9. T. C. Hollocher and M. M. Weber, Nature, 195(4838):247 (1962).
10. V. B. Golubev, Yaguzhinskii, and Volkov, Biofizika (in press).
11. T. H. James, J. M. Suell, and A. Weissberger, J. Am. Chem. Soc., 60(98):2084 (1938).
12. Hironobu Kawasaki, J. Chem. Soc. Japan, Industr. Chem. Sect., 67(10):1554 (1964).
13. A. G. Chetverikov, L. A. Blyumenfel'd, and G. V. Fomin, Biofizika, 10:477 (1965).
14. I. A. Kazarnovskii, N. P. Lipikhin, and M. V. Tikhomirov, Dokl. Akad. Nauk SSSR, 120:1038 (1958).
15. W. Schamb, C. Setterfield, and R. Wentworth, Hydrogen Peroxide [Russian translation], IL, Moscow (1958).
16. A. E. Kalmanson, I. G. Kharitonenkov, A. G. Chetverikov, and L. A. Blyumenfel'd, Biofizika, 8:722 (1963).
17. I. G. Kharitonenkov, Candidate's Dissertation [in Russian], Moscow State University (1965).
18. G. V. Fomin, L. G. Ignat'eva, and L. A. Blyumenfel'd, Biofizika (in press).
19. F. Horowitz, Protein Chemistry and Function [Russian translation], Izd. Mir, Moscow (1965).

PHOTOREDUCTION OF WEAK ORGANIC ACCEPTORS BY WATER (OH⁻ IONS)

G. V. Fomin and V. A. Rimskaya

Institute of Chemical Physics
Academy of Sciences of the USSR

In previous articles [1, 2], we considered the single-electron reduction of such organic electron acceptors as tetracyanethylene, chloranil, bromanil, n-benzoquinone, naphthoquinone, tetracyanoquinodimethane, and methylene blue during their reaction with water. Reduction of these acceptors (to form the corresponding ion-radicals) can result either from direct electron transfer from an OH⁻ ion to the acceptor [1] or transfer proceeding through formation of some intermediate electron donor, as is presumed for n-benzoquinone and naphthoquinone [3-5].

We investigated the analogous reaction with weak electron acceptors, such as anthraquinone and its derivatives. It was found that, in contrast to the aforementioned acceptors, no corresponding ion-radicals are formed during solution of anthraquinones in aqueous and water-alcohol alkali solutions.

The absence of the reaction characteristic of many acceptors is apparently due to the low electron affinity of anthraquinones.

It can be assumed that the electron affinity of the acceptor increases in the excited state, making the reaction possible.

The present work was devoted principally to investigation of the kinetic regularities in and mechanism of this photochemical reaction.

EXPERIMENTAL METHOD

The EPR signals produced during photoirradiation of acceptors were recorded with a standard EPR-2 IKhF radiospectrometer. The specimens were irradiated in glass ampules 1 mm in diameter in the radiospectrometer resonator, using a DKSSh-1000 bulb (full-spectrum light or through interference filters with a pass band of 8-10 mμ). The portion of solution irradiated was about 0.01 ml. All the measurements were made with a previously calibrated internal standard (Mn^{2+} ions in an MgO crystal); the experimental values on most of the graphs were then plotted in terms of A_{sign}/A_{st} (the amplitude ratio of the EPR signals of the ion-radical and internal standard), which was proportional to the radical concentration. The proportionality factor between A_{sign}/A_{st} and the concentration of 2,7-CO_3^--anthraquinone radicals, for which the main kinetic measurements were made, was $7 \cdot 10^{14}$. The signal form remained

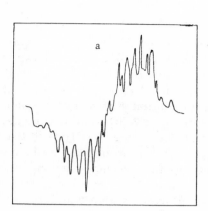 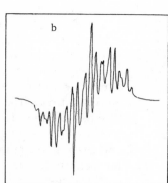 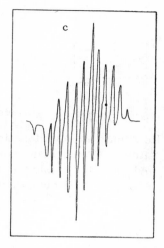

Fig. 1. EPR spectra of ion-radicals. a) 2-SO$_3^-$-anthraquinone; b) 2,6-SO$_3^-$-anthraquinone; c) 2,7-SO$_3^-$-anthraquinone.

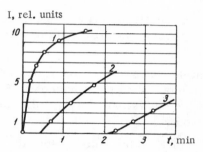

Fig. 2. Kinetic curves for accumulation of 2,7-SO$_3^-$-anthraquinone radicals during photoirradiation. 1) Without oxygen; 2) saturation with air; 3) saturation with oxygen; [A] = 2 · 10^{-3} mole/liter; [OH$^-$] = 5 · 10^{-2} mole/liter.

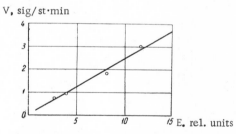

Fig. 3. Reaction rate as a function of light intensity. [A] = 2 · 10^{-3} mole/liter; [OH] = 5 · 10^{-2} mole/liter; 440-mμ filter.

constant in all the kinetic measurements. The signal and standard spectra were recorded at sweep rates of about 100 Oe/min and a sweep range of 50 Oe. This enabled us to record the signals every 15-20 sec during the kinetic measurements.

In the temperature measurements, the specimen in the radiospectrometer resonator was heated in a stream of air at a predetermined temperature.

We employed a calibrated photocell graciously furnished by A. Yu. Borisov (Moscow State University, Biology Faculty) to determine the number of quanta incident on the specimen in establishing the quantum yield of the reaction. The oxygen was removed from the solutions before irradiation and they were then saturated with argon (except in certain special experiments). Our main experiments were conducted with water-soluble sulfur-substituted anthraquinones. This enabled us to perform experiments in aqueous solutions and prevented radical formation as a result of the known photooxidation of many organic solvents by detachment of hydrogen from a saturated carbon atom [6, 9].

EXPERIMENTAL RESULTS AND DISCUSSION

Investigation of Kinetic

Regularities of Reaction

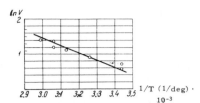

Fig. 4. Reaction rate as a function of temperature on Arrhenian coordinates. $[A] = 2 \cdot 10^{-3}$ mole/liter; $[OH] = 5 \cdot 10^{-2}$ mole/liter; 440-mμ filter.

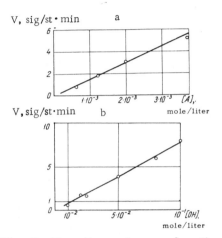

Fig. 5. Reaction rate as a function of concentration. a) 2,7-SO$_3^-$-anthraquinone at $[OH^-] = 5 \cdot 10^{-2}$ mole per liter; b) OH$^-$ ions at $[A] = 2 \cdot 10^{-3}$ mole/liter; 440-mμ filter.

The EPR spectra produced during photoirradiation of the anthraquinones investigated are shown in Fig. 1. It must be noted that the reaction occurred only at $\lambda > 500$ mμ. Figure 2 (1) is a typical kinetic curve for radical accumulation during irradiation of an Ar-saturated solution of 2,7-SO$_3^-$-anthraquinone. Our experimental conditions were usually such that $[OH^-] \gg [A]$ (where A is the anthraquinone molecule) and the kinetic curve was for a pseudo-monomolecular reaction. The curves in Fig. 2 were obtained with the test solution saturated with air and oxygen and will be discussed below.

We made standard kinetic measurements of the initial reaction rate as a function of light intensity, temperature, and reagent concentrations. Figures 3-5 present the results of these measurements.

The reaction was first-order with respect to both light intensity (being single-quantum) and components.

The activation energy of the reaction, determined from the temperature measurements (Fig. 4), corresponded to the activation energy of diffusion, $E_a = 2$-3 kcal/mole.

Such values of E_a often indicate that the photochemical reaction being studied is elementary and not associated with intermediate or secondary dark processes.

The following reaction pattern can be deduced from the results of our measurements:

$$A + h\nu \rightarrow A^\bullet$$
$$A^\bullet + OH^- \rightarrow A^- + OH. \tag{I}$$

It might also be surmised that ion-radical formation results from detachment of a hydrogen atom from an excited anthraquinone molecule when it collides with an unexcited molecule. However, detachment of a hydrogen atom from an aromatic carbon atom is inefficient [6]; moreover, we should have found the reaction to be second-order with respect to anthraquinone concentration in this case. It might be assumed that a hydrogen atom is detached from a water molecule. In this case, however, we would have a zero-order reaction with respect to OH$^-$ concentration.

These suggested patterns are thus refuted by our experiments. It must be noted that no anthraquinone radicals were formed at pH's above 11. This was apparently due, however, not to a decrease in the radical-formation rate but to a sharp increase in the radical-decay rate as a result of the protonization of the radicals at such pH's and a subsequent disproportionation

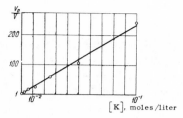

Fig. 6. Dependence of V_0/V on concentration kI (explanation in text). $[A] = 2 \cdot 10^{-3}$ mole/liter; $[OH^-] = 5 \cdot 10^{-2}$ mole/liter.

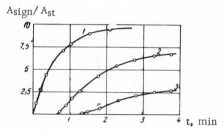

Fig. 7. Kinetic curves at different $K_3Fe(CN)_6$ concentrations in 2,7-SO_3^--anthraquinone solution. $[A] = 2 \cdot 10^{-3}$ mole/liter; $[OH^-] = 5 \cdot 10^{-2}$ mole/liter. $[K_3Fe(CN)_6]$- 0 mole per liter (1); $2 \cdot 10^{-4}$ mole/liter (2); $4 \cdot 10^{-4}$ mole/liter (3).

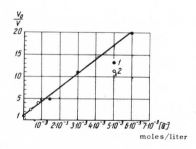

Fig. 8. Dependence of V_0/V on $K_4Fe(CN)_6$ concentration. 1) For $K_4Fe(CN)_6$; 2) for $K_3Fe(CN)_6$; the rate was measured after the induction period; $[A] = 2 \cdot 10^{-3}$ mole per liter; $[OH^-] = 5 \cdot 10^{-2}$ mole/liter.

reaction (the rate constant of disproportionation of semiquinone radicals, such as duroquinone [6], is more than 10^2 times as large as the corresponding constant for ion-radicals).

We can thus evaluate the maximum radical-dissociation constant. Actually, if the reaction is carried out in alcoholic alkali, radical formation is governed principally by detachment of hydrogens from the alcohol and is independent of the pH.

However, no radicals are formed during photoirradiation at the hydroxyl-ion concentration corresponding to pH 11. It can thus be assumed that the dissociation constant of the radical pK $\ll$ 11.

Determination of Nature of Excited State and Quantum Yield of Reaction

We were interested in evaluating the lifetime of the excited state participating in the photochemical reaction. For this purpose, we conducted experiments with known excitation-damping agents, such as the anions I^-, CNS^-, $S_2O_3^{2-}$, and $Fe(CN)_6^{4-}$, in the solution. We found that the action of all the aforementioned damping agents conformed well to the Stern−Volmer equation [8]:

$$\frac{v_0}{v} = 1 + k \tau [Q^-],$$

where v_0 is the reaction rate in the absence of the damping agent, v is the reaction rate in the presence of the damping agent, k is the rate constant of the damping reaction, τ is the lifetime of the excited state, and $[Q^-]$ is the damping-agent concentration.

The value of $k\tau$, found from the appropriate experimental functions, was 2-3 $\cdot 10^3$ for different damping agents. For example, Fig. 6 shows the results of measurements with I^- as the damping agent. If we assume that the damping is controlled by diffusion and $k = 10^{10}$ liters/mole $\cdot$ sec, the lifetime of the excited state $\tau = 10^{-7}$ sec and a triplet excited state evidently participates in the reaction.

We evaluated the quantum yield of the reaction (at a wavelength of 440 mμ), where η is the number of quanta absorbed divided by the number of radicals formed.

We determined the number of quanta absorbed and the radical yield after photoirradiation for 30 sec; η was found to be about 1. It should not be surprising that a triplet excited state participates in the reaction or that the quantum yield is simultaneously high, since the transition from a singlet to a triplet excited state is known to be highly efficient for many aromatic ketones [7].

Anthraquinone-Sensitized Reduction of

Weak Electron Acceptors by Water (OH$^-$)

It can be seen from Fig. 2 that an induction period appears in the kinetic curve when the solution contains oxygen. It is also known that anthraquinone ion-radicals are oxidized by oxygen to regenerate the anthraquinone molecule and produce hydrogen peroxide [9]. It can be assumed that electron transfer from water to oxygen through intermediate formation of anthraquinone ion-radicals occurs during the induction period. Such active compounds as HO_2 and H_2O_2 are formed in the system and can participate in secondary reactions, complicating the processes taking place in the reaction mixture. We therefore studied sensitized reduction, employing the Fe^{3+} ion (in the $[Fe(CN)_6]^{3-}$ complex) as the acceptor. Figure 7 shows the corresponding kinetic curves for two $K_3Fe(CN)_6$ concentrations.

The induction period was an approximately linear function of the $[Fe(CN)_6]^{3-}$ concentration over a broad concentration range. Qualitative analysis confirmed that $[Fe(CN)_6]^{4-}$ was formed in the reaction mixture after the reaction.

Almost all the $[Fe(CN)_6]^{3-}$ ions were evidently reduced during the induction period.

If we construct a Stern−Volmer function for the reaction rate after the induction period, plotting the initial $[Fe(CN)_6]^{3-}$ concentration along the abscissa, it closely follows the line representing the Stern−Volmer function for damping of excited $[Fe(CN)_6]^{4-}$ ions (Fig. 8); it can thus be assumed that the retardation of the reaction rate after the induction period as the initial $K_3Fe(CN)_6$ concentration rose was due to damping of the reactive excited states of the $[Fe(CN)_6]^{4-}$ ions formed.

Thus, using the Fe^{3+} ion as an example, we can assume the following pattern for anthraquinone-sensitized reduction by OH$^-$ ions:

$$OH^- + A \xrightarrow{h\nu} OH + A^-$$
$$\underline{A^- + Fe^{3+} \rightarrow A + Fe^{2+}}$$
$$OH^- + Fe^{3+} \xrightarrow{h\nu} OH + Fe^{2+} \tag{II}$$

In all these reactions, it is also necessary to take into account the secondary decay of semiquinone radicals and, possibly, of anthraquinone radicals as well. Active OH$^-$ radicals are actually formed during the reaction; these can, for example, be irreversibly attached to the acceptor by a double bond. Disproportionation of the OH radicals can yield atomic hydrogen, whose reaction also leads to irreversible decomposition of the acceptor; furthermore, molecular oxygen causes reversible breakdown of the radicals, oxidizing them to the initial anthraquinone. However, this reaction entails formation of active HO_2 radicals, which can result in irreversible decomposition of anthraquinones, and hydrogen peroxide, which oxidizes semiquinone radicals, decomposing them to active OH radicals and OH$^-$. It must be noted that if consideration is given to the disproportionation of OH radicals to form O_2, and to the oxidation of semiquinone radicals by molecular oxygen and hydrogen peroxide, the overall stoichiometry is the same as that of system I.

Participation of the aforementioned reactions may explain the possibility of a 100% radical yield in the system studied and the substantial decrease in radical yield when reaction II has a long induction period.

LITERATURE CITED

1.   G. V. Fomin, L. A. Blyumenfel'd, and B. I. Sukhorukov, Dokl. Akad. Nauk SSSR, 157(5): 1199 (1964).

2.   G. V. Fomin, L. A. Blyumenfel'd, R. M. Davydov, and L. G. Ignat'eva, present volume, p. 90.
3.   M. Eigen and R. Matthies, Chem. Ber., 94:3309 (1961).
4.   T. C. Hollocher and M. M. Weber, Nature, 195(4838):247 (1962).
5.   V. B. Golubev, L. A. Yaguzhinskii, and A. A. Volkov, Biofizika (in press).
6.   Bridge and G. Porter, Proc. Roy. Soc., A244:259 (1958).
7.   F. Wilkinson, J. Phys. Chem., 66(12):2569 (1963).
8.   B. I. Stepanov, Luminescence of Complex Molecules [in Russian], Izd. AN BSSR (1955).
9.   C. F. Wells, Trans. Faraday Soc., Vol. 57, No. 10 (1961).

THERMODYNAMICS AND ROLE OF WATER
IN AUTOOSCILLATORY PROCESSES

Yu. P. Syrnikov

S. M. Kirov Leningrad Academy of Lumber Technology

The importance of autooscillatory processes in living organisms is now well known. These processes occur at the most diverse levels. Autooscillations at the "molecular" or "thermodynamic" level have been the most thoroughly studied and are perhaps the most significant for interpreting intracellular processes. A number of experiments have demonstrated that such autooscillations can occur in both thermodynamically nonequilibrated and thermodynamically equilibrated systems. Thus, Shnol' [1] has established that there are periodic changes in ATPase activity and SH-group titer in actin, myosin, and actomyosin solutions at constant temperature and pressure. This indicates that some sort of autooscillatory process takes place in the system, in all probability being associated with conformational changes in the molecules. The explanation proposed by Shnol' presumes that the conformational changes are synchronous, synchronization being achieved by transmission of hydrophilic—hydrophobic "waves," which propagate in the water filling the space between the protein molecules. We will show below that there is a more natural and more general explanation for the results obtained.

A general examination of the thermodynamics of processes similar to those detected by Shnol' and of the role of water in such processes is of interest. As analysis has shown, the problem can be attacked through the concepts of the thermodynamics of small systems. In the thermodynamics of small systems, the thermodynamic properties and behavior of a small system depend both on the usual thermodynamic parameters and on the composition of the medium [2]. This results from the fact that the intermolecular interaction between the system and medium cannot be neglected for a small system.

A macromolecular solution can be regarded as an ensemble of small systems, with each macromolecule being treated as a system in itself. These macromolecules can be in different states, with transitions between them. The best known transitions are of the helix-globule type. They can be regarded as intramolecular melting, or melting of a small system. One characteristic feature of the phase transitions in small systems is their smoothness, i.e., the lack of any abrupt jump in thermodynamic functions [2]. If small systems are immersed in a solvent, the phase-transition conditions are governed both by the temperature and pressure and by the solvent composition, i.e., the character of the medium. If the solvent composition in turn depends on the state of the macromolecules, a type of feedback and autooscillation can develop. Such situations should arise especially readily in aqueous solutions, since the solute affects the water structure; this effect extends over very large areas.

Let us consider the following model. We have an ensemble of small systems immersed in a two-structure solvent. The state of a small system depends on its environment, e.g., on the solvent composition. The solvent composition in turn depends on some parameter characterizing the ensemble of systems; for example, it depends on the ratio of the numbers of systems with different conformations if conformational transitions occur. We must analyze the conditions under which cyclic changes in the ratio of the numbers of systems in different conformational (or phase) states can take place without disrupting the overall thermodynamic equilibrium of the larger system, e.g., the ensemble of small systems plus the solvent. It must incidentally be noted that explanation of the periodic change in ATPase activity and SH-group titer does not require us to assume synchroneity of the transitions in all the macromolecules; it is sufficient for only the ratio of the numbers of systems in different states to change.

It is therefore also unnecessary to assume hydrophilic−hydrophobic waves. We will employ the two-state approximation to describe the conformational transitions. This approximation is quite well founded for phase transitions of the first type in small-system ensembles and can be applied to transitions of the helix-globule type.

Let us write an equation for the thermodynamic potential of the larger system:

$$\Phi_\delta = N_s \left[c\mu_1 (p,\ T,\ x,\ c,\ b) + (1-c)\mu_2 \right.$$

$$(p,\ T,\ x,\ c,\ b)] + \omega N \left[b\mu_A (p,\ T,\ x,\ c) + (1-b) \times \mu_B (p,\ T,\ x,\ c) \right] - kT \ln \frac{w!}{w_A!\ w_B!} + \omega kT \ln x. \qquad (1)$$

Here, N_S is the total number of solvent molecules, c is the concentration of the first solvent structure, μ_1 and μ_2 are the chemical potentials of the first and second solvent structures, respectively; ω is the total number of macromolecules; μ_A and μ_B are the numbers of macromolecules in states A and B; b is the relative concentration of macromolecules in state A; x is the macromolecule concentration in the solution; N is the number of units in the macromolecular chain; and μ_A and μ_B are the chemical potentials per unit for macromolecules in states A and B, respectively. It must be noted that the chemical potentials μ_A and μ_B are defined as the ratio of the thermodynamic potential of the entire molecule to the number of units and not as the derivative of this potential with respect to the number of units. These two definitions give different values for the chemical potential.

Then the variation in the thermodynamic potential of the larger system should be zero when it is in thermodynamic equilibrium at constant temperature and pressure. In addition to p and T, our formulation contains the constants N_S, ω, N, and x. Thus, only c and b vary. By varying Φ_δ and setting the variation to zero, we obtain the two relationships

$$N_s \left[c\left(\frac{\delta\mu_1}{\delta b}\right) + (1-c)\left(\frac{\delta\mu_2}{\delta b}\right) \right] + \omega \left[N\Delta\mu + kT \ln\frac{b}{1-b} \right] = 0, \qquad (2)$$

$$\omega N \left[b\left(\frac{\delta\mu_A}{\delta c}\right) + (1-b)\left(\frac{\delta\mu_B}{\delta c}\right) \right] + N_s \left[\Delta\mu_s + c\left(\frac{\delta\mu_1}{\delta c}\right) + (1-c)\left(\frac{\delta\mu_2}{\delta c}\right) \right], \qquad (3)$$

where $\Delta\mu = \mu_A - \mu_B$ and $\Delta\mu_S = \mu_1 - \mu_2$; the derivatives are taken at constant p, T, N_S, N, and x.

If these relationships are not identities, we have two equations and the equilibrium in the larger system corresponds to fully defined values of c and b. If relationships (2) and (3) do not constitute a complete equation system, i.e., if one or both relationships are identities, a set of values of c and b is compatible with thermodynamic equilibrium in the larger system and cyclic changes can occur in these factors. In order for the relationships to be identities, it is sufficient that the expressions in brackets equal zero over a certain range of variation in c and b.

The chemical potentials of the solvent structures and μ_A and μ_B must be definite functions of concentration for this to occur. Let us analyze the necessary conditions. Relationship (2) is obtained by setting the coefficient of the variation δb at zero, while relationship (3) is obtained by setting the coefficient of the variation δc at zero. Equation (2) must therefore be an identity for b, while relationship (3) is an identity for c. When the second expression in brackets in Eq. (2) equals zero, we have the equilibrium condition for transitions from state A to state B

$$\ln \frac{b}{1-b} = -\frac{N\Delta\mu}{kT}, \tag{4}$$

when the first term in brackets equals zero we have the relationship

$$\frac{c}{1-c} = -\frac{\left(\dfrac{\partial\mu_2}{\partial b}\right)}{\left(\dfrac{\partial\mu_1}{\partial b}\right)}. \tag{5}$$

Thus, if these relationships are satisfied over a certain range of variation in b, Eq. (2) becomes an identity and a set of values of c and b is compatible with thermodynamic equilibrium of the larger system. Analyzing relationship (3), we similarly obtain a condition for converting it to an identity over a range of c.

$$\Delta\mu_s + c\left(\frac{\partial\mu_1}{\partial c}\right) + (1-c)\left(\frac{\partial\mu_2}{\partial c}\right) = 0 \tag{6}$$

$$\frac{b}{1-b} = -\frac{\left(\dfrac{\partial\mu_B}{\partial c}\right)}{\left(\dfrac{\partial\mu_A}{\partial c}\right)}. \tag{7}$$

The first equality is the condition for equilibrium between the solvent structures, while the second is similar to Eq. (5).

It follows from the foregoing that, if μ_1, μ_2, μ_A, and μ_B are functions of concentration such that Eqs. (4)-(7) are satisfied, cyclic processes can occur in the system without disrupting the thermodynamic equilibrium of the larger system, resembling gigantic cyclic "fluctuations." It must be emphasized that Eqs. (4)-(7) have nontrivial solutions for μ_1, μ_2, μ_A, and μ_B only if $\Delta\mu$ and $\Delta\mu_s$ depend on c and b, respectively, in some definite fashion. In other words, feedback should occur. The feedback results from the smallness of the system and, hence, from the dependence of $\Delta\hat{\mu}$ on c on one hand and from the influence of the macromolecular conformation on the water structure on the other. An autooscillatory process of the following type is then possible. When the concentration b randomly increases, the equilibrium between the water structures is displaced in such fashion that $\Delta\hat{\mu}$ decreases and this leads to a still larger increase in b, etc. A positive feedback process develops, also being manifested in a periodic change in macromolecular conformation. The amplitude of this process must depend in some complex fashion on the concentration and temperature, but the existence of an optimum concentration is rather easily explained. The influence of the macromolecules on the water structure is relatively small at low concentrations; little "two-structure" water remains at high macromolecule concentrations and the relative effect of a change in water structure on the conformational equilibrium is small. There is thus an optimum concentration at which the oscillation amplitude is greatest. The oscillation period should be governed by the kinetics of the conformational changes.

LITERATURE CITED

1.   S. E. Shnol', in: Molecular Biophysics [in Russian], Izd. Nauka, Moscow (1965).
2.   T. L. Hill, Thermodynamics of Small Systems, Benjamin, New York (1963).

STRUCTURED WATER ADJOINING THE SURFACE LAYER OF ERYTHROCYTES

K. S. Trincher

Institute of Biological Physics
Academy of Sciences of the USSR

The surface layer of erythrocytes washed free of serum proteins and suspended in physiological solution has a certain organizing effect on the adjacent extracellular water, giving it the properties of structured water.

In studying the injurious action of various agents on erythrocyte suspensions, we found that the water in contact with the cell surface had protective properties by virtue of its structuring. However, if the erythrocytes were incubated in physiological solution previously subjected to magnetization, the water itself became an injurious agent, evidently having acquired unusual structural properties in the constant magnetic field.

Our experiments with different injurious chemical and physical agents indicate that structured extracellular water plays a material role in stabilizing the surface layer of erythrocytes.

EXPERIMENTAL METHOD AND RESULTS

Effect on Erythrocytes of an Injurious Chemical

Agent: High OH-Ion Concentration

Twice-washed erythrocytes from fresh rat blood were suspended in physiological solution at dilutions of from 350 to 3500 times with respect to their concentration in the blood. Figure 1 shows the optical density (D) of the erythrocyte suspensions, measured in an FEK-M2 photoelectrocolorimeter with a blue filter, as a function of the erythrocyte concentration (curve 1) and of the corresponding hemoglobin concentration (curve 2) after completion of hemolysis in the seven specimens investigated. Hemolysis was carried out by adding 5 ml of isotonic buffer solution at a pH of about 9.97 to 1 ml of suspension [1-3].

As can be seen, the optical density was a strictly linear function of concentration at high erythrocyte (and thus hemoglobin) dilutions. Extrapolating the curves to zero concentration, we found the background optical density D_b to be 0.07.

Figure 2 shows the kinetics of hemolysis under the action of an alkaline medium for seven erythrocyte suspensions with different concentrations. The experiments were conducted at 32°C $\pm$ 0.2° and the value of D, determined in parallel experiments, ranged from 0 to 2%. We performed five series of experiments, using fresh rat blood each time.

As can be seen, the rapidity with which the erythrocytes were decomposed under the action of the alkaline buffer increased as the initial erythrocyte concentration decreased.

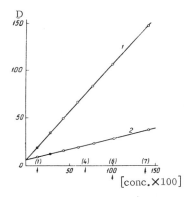

Fig. 1. Optical density as a function of erythrocyte concentration (1) and hemoglobin concentration (2). Optical density (D) is shown along the ordinate; erythrocyte concentration (C) is shown along the abscissa. Erythrocyte concentration in the blood is taken as one.

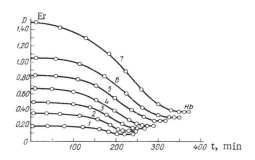

Fig. 2. Kinetics of hemolysis for different erythrocyte concentrations. Time of exposure to isotonic buffer, in min, is shown along the abscissa. 1–7) Different erythrocyte concentrations.

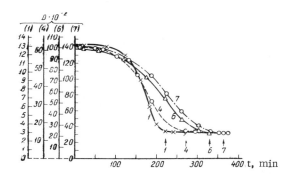

Fig. 3. Kinetics of hemolysis for different erythrocyte concentrations, with different time scales along ordinate for curves 1, 4, 6, 7 (symbols as in Fig. 2).

In order to make a more precise determination of the end of the hemolysis process, we took four of the specimens shown in Fig. 2 and plotted a family of curves for them, varying the time scale in an appropriate manner.

Figure 3 shows the curves thus obtained, with the background optical density (D_b = 0.07) subtracted. The arrows indicate the points at which hemolysis terminated. Figure 4 shows the time at which hemolysis began in the alkaline medium as a function of erythrocyte concentration; as can be seen, the time required for hemolysis decreased with the erythrocyte concentration. Extrapolation to zero concentration yields the minimum time for breakdown of one erythrocyte surrounded by a cell-free alkaline medium.

The function found for the erythrocyte-decomposition time in isotonic alkaline buffer indicates that erythrocytes have a mutual protective effect, which is obviously exerted through the intercellular aqueous medium. It can be surmised that the surface layer of the erythrocytes orders the structure of the adjacent water and that the ordered water has a protective effect, which increases with the cell concentration.

Effect of Cs^{137} γ-Rays in a Dose of 5 kR on Erythrocytes

Erythrocytes from fresh rat blood were subjected to γ-irradiation under two different sets of conditions: a) suspended in physiological solution; b) in a "densely packed" state produced by centrifugation at 3000 rpm for 5 min. Figure 5 shows the results of one experiment conducted by isotonic alkaline hemolysis: curve 3 represents the control (an erythrocyte suspension not subjected to irradiation), curve 1 represents an experiment in which the erythrocytes were irradiated in suspension, and curve 2 represents an experiment in which the erythrocytes were densely packed at the bottom of the test tube. As can be seen, the erythrocytes irradiated while densely packed exhibited about 20% greater protection against the action of penetrating radiation than the suspended erythrocytes. This result can be attributed to the fact that the water immediately adjacent to the erythrocyte surface had the properties of structured water, which provides protection. The structuring of the water should be greater for densely packed erythrocytes than for suspended erythrocytes [4].

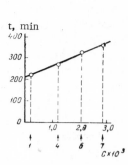

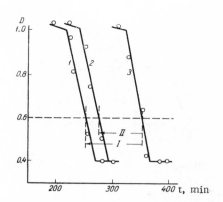

Fig. 4. Hemolysis time in alkaline medium as a function of erythrocyte concentration.

Fig. 5. Optical density (D) as a function of time for which erythrocyte suspension was exposed to alkaline medium. 1) Erythrocytes irradiated in suspension; 2) erythrocytes irradiated in "densely packed" form; 3) control.

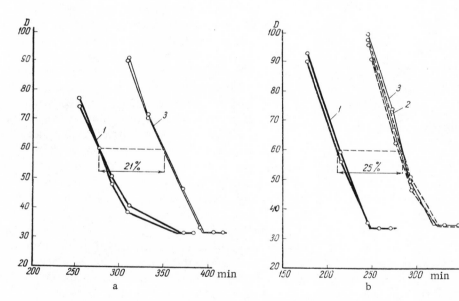

Fig. 6. Ordinate the same as in Fig. 5. a: 1) Erythrocytes incubated in physiological solution subjected to magnetization. b: 2) Erythrocytes incubated in physiological solution boiled and cooled after magnetization; 3) control.

Effect on Erythrocytes of Water

Subjected to Magnetization

Magnetization of the water used in boilers has been employed in thermotechnics for a number of years. If the water is passed through a strong constant magnetic field immediately before use, an amorphous, easily removed scale is formed during evaporation rather than the hard crystalline layer characteristic of unmagnetized water. No theory has as yet been developed to account for this phenomenon, i.e., precipitation of the salts dissolved in magnetized water in the form of a gel-like layer during evaporation [5-11].

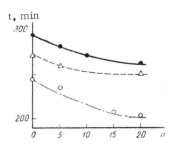

Fig. 7. Hemolysis time as a function of number of times physiological solution was passed through constant magnetic field. Results of three different experiments.

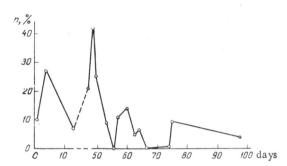

Fig. 8. Damage to erythrocytes (% of control) incubated in magnetized physiological solution as a function of experiment duration.

In our experiments, we passed physiological solution through a glass tube, a small portion of which was in the field of a permanent magnet with a strength of 4000 Oe. The solution dripped from the lower end of the tube, was collected in a container, and was again passed through the tube. This procedure was repeated 20 times. Fresh rat blood was added to the magnetized physiological solution in a ratio of 1:100. The resultant erythrocyte suspension was incubated at room temperature for 15-18 h. The control was a similar erythrocyte suspension prepared from the same portion of blood in unmagnetized physiological solution.

Our subsequent investigations were conducted by isotonic alkali hemolysis.

Figure 6a shows two control (3) and two experimental (1, 2) curves representing the kinetics of erythrocyte breakdown in an alkaline solution. As can be seen, the experimental curves are displaced to the left by about 21% with respect to the control curves. This means that the magnetized physiological solution in which the erythrocytes were incubated had an injurious effect on them. The damaged erythrocytes were hemolyzed substantially more rapidly in the alkaline medium than erythrocytes from the same blood incubated in physiological solution not subjected to magnetization. Figure 6b presents two curves obtained in experiments where the magnetized physiological solution was boiled and then cooled (1, 2). As can be seen, while the unboiled magnetized physiological solution damaged a substantial number of the erythrocytes (about 25%), the boiled solution had no injurious effect at all.

Figure 7 shows the saturation point of magnetization of the physiological solution with respect to the extent of the erythrocyte damage. The time required for hemolysis to occur is plotted along the ordinate, while the number of times the physiological solution was passed through the magnetic field is plotted along the abscissa. As can be seen from the curves for three different experiments with erythrocytes from the blood of different rats, passage of the physiological solution through the magnetic field 20 times caused it to have the maximum injurious effect on the erythrocytes.

Our experiments were begun in Vladivostok in May, 1965 and continued in Moscow from June to September, 1965 [12]. Figure 8 shows the results of all the experiments performed in Moscow. As can be seen, the damage to the erythrocytes caused by exposure to magnetized physiological solution, expressed as a percentage of the control figure, exhibited irregular fluctuations. The injurious effect was greatest at the beginning of the summer and reached more than 40% in one experiment. There was then a decrease in the effect: no damage was detected in three experiments and the effect ranged from 3 to 5% from September, 1965 onward. We discontinued the investigation at this time.

The phenomenon observed (the irregular injurious effect of magnetized water on erythrocytes) has still not been explained, particularly since there are at present no theories to account for the structural changes produced in water by magnetization. However, in our opinion, the thermodynamic approach to the study of water structure enables us to advance a few hypotheses regarding this effect. Water is an aggregate of two continuously interconvertible microphases: one consists of structured or crystalline water and the other of unstructured liquid water. The internal energy U of water at room temperature is:

$$U = \int_0^{300} S_{stand} dT = 16.7 \cdot 300 \simeq 5 \; kcal/mole,$$

where S_{stand} is the standard entropy of water (T = 293) and equals about 16.7 cal $\cdot$ deg^{-1} $\cdot$ mole^{-1}, where T is the absolute temperature.

According to Pauling [13], the energy of a hydrogen bond $E_{H...O}$ is about 4.5 kcal $\cdot$ mole^{-1}.

Hence, U > $E_{H...O}$ and all the water should theoretically be in the liquid state, containing no crystalline−water microphases. Actually, however, liquid water does contain such microphases, which are stabilized by hydrogen bonds. This contradiction is resolved by the fact that the energy of hydrogen bonds is locally fixed in the crystalline structures, while the thermal energy fluctuates. The latter energy is distributed nonuniformly over the molecules of the liquid microphases. The fluctuations in thermal energy in water result in continuous phase transitions between the crystalline and liquid microphases, these not requiring energy exchange with the ambient medium. In other words, microphase transitions between crystal and liquid take place in water without the participation of an external heat source or sink [14].

On the average, the microphase transitions in water should naturally take place in either direction with equal probability.

Two hypotheses can be advanced to account for the irregularity of the structural changes in water under the action of a constant magnetic field.

The first hypothesis holds that, since the microphase transitions in water do not generally require expenditure of energy, some negligibly small external physical factor, whose nature is unclear, can cause preferential formation of one microphase or of an unusual combination of the two microphases. The observed irregular effect of magnetization of water may be due to the fact that an irregularly arising external factor produces an unusual water structure, which is stabilized under the simultaneous action of a constant magnetic field.

The second hypothesis holds that since the microphase transitions in water do not require energy, there can be states involving preferential formation of one phase or an unusual combination of the two microphases. It is likely that the observed irregular effect of magnetization consists in stabilization of the microstates by a constant magnetic field; there can also be a deviation from the most probable combination of crystalline and liquid microphases.

LITERATURE CITED

1. K. S. Trincher, Biofizika, 4(2):79 (1958); 4:731 (1959).
2. A. M. Kuzin and K. S. Trincher, Biofizika, 5:533 (1960).
3. K. S. Trincher and L. V. Orlova, Biofizika, 10(3):540 (1965).
4. K. S. Trincher and A. P. Kuzin, Radiobiologiya, 4:36 (1964).
5. T. Vermeiren, Belgian Patents 460560 and 4894997 (1945).
6. N. P. Lodozhishkina, Teploenergetika, 11:45 (1959).
7. G. A. Rogal'-Levitskii, Vodosnabzhenie i Santekhnika, No. 2 (1961).
8. E. A. Friedel, Chem. Ztg., 84:16 (1960).

9. I. L. Eindel'shtein (editor), Magnetic Treatment of Water [in Russian], Alma-Ata (1961).

10. V. I. Pinenko and S. M. Parov, Teploenergetika, 9:63 (1962).

11. V. I. Klassen and S. V. Tserbakova, Gornyi Zh., 5:58 (1965).

12. I. I. Brekhman, Yu. S. Malinin, K. S. Trincher, Oscillatory Processes in Biological and Chemical Systems (in press).

13. L. Pauling, The Nature of the Chemical Bond, 3rd ed., New York (1960), p. 450.

14. K. S. Trincher, Uspekhi Sovr. Biol., Vol. 3 (1966).